T0186905

Simple Excellence

Organizing and Aligning the
Management Team in a
Lean Transformation

Simple Excellence

Organizing and Aligning the Management Team in a Lean Transformation

Adam Zak and Bill Waddell

CRC Press
Taylor & Francis Group
Boca Raton London New York

CRC Press is an imprint of the
Taylor & Francis Group, an **informa** business

A PRODUCTIVITY PRESS BOOK

Productivity Press
Taylor & Francis Group
270 Madison Avenue
New York, NY 10016

© 2011 by Taylor and Francis Group, LLC
Productivity Press is an imprint of Taylor & Francis Group, an Informa business

No claim to original U.S. Government works

Printed in the United States of America on acid-free paper
10 9 8 7 6 5 4 3 2 1

International Standard Book Number: 978-1-4398-3845-7 (Hardback)

Library of Congress Cataloging-in-Publication Data

Zak, Adam, 1952-
    Simple excellence : organizing and aligning the management team in a lean transformation / Adam Zak, Bill Waddell.
        p. cm.
    Includes index.
    ISBN 978-1-4398-3845-7 (hardcover : alk. paper)
    1. Organizational effectiveness. 2. Teams in the workplace. 3. Management. I. Waddell, Bill. II. Title.

HD58.9.Z347 2011
658.4'022--dc22   2010038206

Visit the Taylor & Francis Web site at
http://www.taylorandfrancis.com

and the Productivity Press Web site at
http://www.productivitypress.com

Contents

Preface

For over 25 years we have been working in and around manufacturing. Our careers began before Jim Womack, Dan Jones, and Dan Roos coined the term *Lean manufacturing* in their watershed book, *The Machine that Changed the World*, and we have seen the global manufacturing landscape transform as companies have worked to understand and apply the principles of Lean thinking. We would like to think that we have even influenced the evolution of Lean a bit through our work and our writing.

We have seen activity-based costing come into and pass out of vogue and the theory of constraints become one of the basic principles of manufacturing. We have witnessed manufacturing quality management evolve as a result of the ideas behind statistical process control (SPC), Six Sigma, and Deming's plan–do–check–act (PDCA) principles. Technology—especially information and communications technology—has enormously impacted manufacturing over the course of our careers. And, of course, we have observed the effects of globalization.

Throughout this period of dramatic change and under the onslaught of new ideas, we have had a chance to observe some very successful companies. Unfortunately, during this time too many great manufacturing companies have dwindled or died out all together. Most companies are somewhere in between, trying to figure it all out. This book is written for the management of those companies. Having had a chance to see several very successful manufacturing companies in action, we have had the great fortune of learning some very important lessons. Our purpose here is to share them.

A couple of overarching principles are woven throughout this book. They keep coming up because they are common among all of the very well-managed companies we have studied and with which we have worked, and strike us as central to their success. First among them is the degree to which the entire management team is engaged and involved. While most companies are relatively compartmentalized—salespeople sell, manufacturing people manufacture, and accounting people account—in the best companies there's a dramatic difference: you need a scorecard to determine who does what even after having spent an hour with the management team. No one's turf is sacred. All are involved in the entire business and work in tightly integrated teams in just about every aspect of the enterprise. This gives them a powerful synergy of knowledge, talents, and ideas.

When we look at the evolution of Lean manufacturing, most of the literature and most of the effort have been aimed exclusively at manufacturing operations (hence *Lean manufacturing*, we suppose). On top of the Lean factory literature, there has been quite a bit aimed at the chief executive officer (CEO) level concerning Lean philosophy and Lean leadership. In recent years accounting, product innovation, and engineering have been slowly added into the mix. In most companies and in most of the literature, however, the rest of the management team (e.g., sales and marketing, human resources, information technology) has been mostly left out of the Lean transformation process. At best, some suggest that sales and marketing recognize that Lean manufacturing has enabled shorter lead-time deliveries or that quality is much better as a result of the Six Sigma effort and that those improvements should have some value in the marketplace. In the excellent companies, sales and marketing are a lot more involved than that. They are the bridge that takes the factory right to the customer's back door.

All of management is completely caught up in the process of driving the company to much higher levels of performance in these great organizations, and the application of Lean and the other emerging principles is hopelessly intertwined with how they do it.

Another common element is simplicity. The best companies are not pursuing more sophisticated computer applications and more detailed financial management tools. They have the common sense to let their personal wisdom, judgment, and experience trump the onslaught of management theories that are often little more than ways to manage through and around complexity that shouldn't be there in the first place. Quite the opposite, in fact—these companies have demonstrated a great talent for peeling away all of the unnecessary clutter and zeroing in on the things that matter. They worry about the value of the products they make, how their customers feel about things, and how well cash is flowing. Anything that clouds their ability to keep a clear eye on those key factors is cast aside.

These companies are as Lean as Lean can be—but they would probably not pass a "Lean audit." They are not hung up on Japanese terminology and do not seem to care very much whether anyone is a certified "black belt." Rather, they have gained a keen appreciation for the principles that underlie Lean manufacturing and Six Sigma and have fashioned the best way to put those principles to work in their businesses.

The companies that have so impressed us are unique, spanning a wide range of markets, industries, and sizes. They are proof that there is no cookie-cutter method of improvement. We have learned from observing them that each company must find its own way along the path to operational excellence. There is no workbook everyone can use, and there are no universal "Ten Steps." What works for one company will not always work for another. That said, there are some common management principles.

The first objective of this book is to clearly lay out those principles. Our secondary aim is to describe how embracing, massaging, and deploying those principles require the commitment of everyone sitting around the table at the staff meeting. No one can stand on the sidelines and keep up "business as usual." The journey to attain the levels of performance of the companies we will cite will be very simple—but it will not be easy.

Authors

Adam Zak is founder and chief executive officer (CEO) of Adam Zak Executive Search and an accomplished senior executive with more than 25 years of experience spanning the areas of management consulting, financial and operations management, and talent acquisition. Adam excels at identifying and attracting business leaders who are experts in Lean continuous improvement, process and operational excellence, and sustainability. His impact on organizations quickly becomes apparent as these talented individuals, in their new leadership roles, consistently go on to make their new companies excellent.

Adam has recruited executives for a diverse mix of Fortune 1000, private-equity, and entrepreneurial companies in the United States and abroad. He has worked in a variety of industries from automotive and aerospace giants implementing sustainable operational excellence cultures to fast-paced consumer product distributors revolutionizing their global supply chains with green management practices.

Earlier in his career Adam earned a bachelor of science degree in accounting from the University of Illinois, became a certified public accountant, and worked with KPMG in both advisory and assurance roles. He subsequently held a variety of financial and general management positions with venture capital, innovative start-up, and *Fortune*-ranked companies in Chicago, Seattle, San Francisco, and Silicon Valley.

Adam has been using his uncommon expertise to help his clients improve their businesses operationally and financially for almost 20 years. He's been a partner with two international executive recruiting firms and now conducts executive searches in a much more creative, entrepreneurial, and results-driven (and Lean) fashion. His unique and highly personalized methodology for recruiting exceptional talent is trademarked as PDCAsearch®. Adam is considered by many to be the most influential leader today in Lean recruiting and Lean executive search.

Recently, Adam's work has focused on helping to improve American health care by bringing operational and process excellence to the health-care industry. He is the innovator behind the recently launched recruiting initiative Crusade for Lean HealthCare Talent.

Adam contributes his time to a number of professional and community organizations. He often speaks and writes on topics including the Lean enterprise, innovation, operational excellence, and sustainability. He is a devoted student of executive leadership, change management, and, most of all, the philosophies and practices of continuous improvement.

Bill Waddell is a consultant, author, and speaker who has spent over 30 years working in just about every facet of manufacturing. Beginning with his education at the University of Cincinnati, Bill came up through manufacturing in virtually every capacity, including accounting, industrial engineering, supply chain, production, and quality management. His resume includes working as manager of advanced manufacturing practices for Emerson's Copeland Division, as vice president of global supply chain for McCulloch, and as vice president of global operations for the Wahl Clipper Corporation. He is widely regarded as one of the leading authorities on Lean manufacturing and Lean enterprise management and strategies.

Bill coauthored the award-winning *Rebirth of American Industry* (with Norman Bodek, PCS Press, 2005) and *Evolving Excellence: Thoughts on Lean Enterprise Leadership* (with Kevin Meyer, iUniverse Inc., 2007). His provocative columns on manufacturing issues on *Evolving Excellence* make it the most widely read manufacturing blog in the world. He is also a regular contributor to *Manufacturing Crunch* and *Pacific Excellence*, and his articles have appeared in the Association for Manufacturing Excellence's (AME's) *Target Magazine*, *APICS Journal*, and a wide variety of other manufacturing publications.

As a consultant, Bill has worked with hundreds of companies on five continents in a wide variety of industries, ranging from multibillion dollar global organizations to his pro bono work with local organizations hiring the handicapped to learn basic manufacturing skills. His focus in recent years has been to work with smaller- to mid-sized companies, often in support of the Manufacturing Extension Partnership programs.

As a speaker and educator, Bill has addressed just about every significant manufacturing form in the world. One of the original "thought leaders" behind the Lean accounting body of knowledge, he has become a regular fixture at the Lean Accounting Summit and at numerous events related to the financial management of manufacturing. He is past technical chair for the International Six Sigma/Lean Quality Conference and has spoken to dozens of groups within the Association for Operations Management

(APICS) on topics related to global supply chain optimization and leading-edge factory scheduling. He has lectured at over a dozen top universities as well as at the U.S. Air Force Institute of Technology on manufacturing strategy, global manufacturing issues, and financial management.

1

A Different Way of Managing

The senior leadership team plods in for that business ritual—the weekly staff meeting. For many of the attendees, this is their only regular communication with each other. In fact, communication is the primary goal of the meeting. The boss leads off with some general announcements: concerns, perhaps, or problems communicated from headquarters or the parent company. Then each manager, in turn, discusses the issues and events within his or her area of responsibility. Some of them may squirm under heat from departmental metrics, having to explain any performance since the last staff meeting that does not meet the benchmark set for labor efficiency, sales levels, or the like.

The sales and marketing manager, always the optimist since that is a requisite character trait, talks about a few minor successes and a few opportunities on the horizon but also expresses a few tactful criticisms of the operating side of the business: high prices because of high costs, customer complaints due to delivery or quality problems, and other obstacles to top-line success beyond the control of the sales and marketing department.

Production follows with a beleaguered explanation of why cost reduction efforts are dragging behind plan, an explanation of problems with the enterprise resource planning (ERP) system that wrought havoc on the factory floor (which the rest of the attendees have trouble following), a thinly veiled complaint about the quality of people being hired by human resources (HR), and an overt plug for a capital investment that is not making the hurdle rate but is needed anyway.

Purchasing goes next and offers up an explanation for late supplier deliveries and warns of impending cost increases due to global commodity market swings. Steel, resins, copper, or fiberboard are going up everywhere, so purchase prices will necessarily increase; the silver lining, however, is that they will affect competitors as well.

The supply chain manager offers little in the way of substantive input. The discussion revolves around inventory, but the supply chain manager is really just the scorekeeper. Inventory is the residual of what was made and what was sold, and the ERP system simply does as it was told, so any inventory excesses or shortages are not the supply chain's responsibility. Sales and marketing wants more inventory—finance wants less, but the supply chain typically disavows much real ownership of the argument since the system drives things.

Human resources and information technology (IT) talk about things in their respective arenas—impending changes in labor laws and a change in the backup schedule that will take the server down at a different time on the weekends—that no one else in the meeting understands or cares much about. Engineering may be on the hot seat for delays and cost overruns in a new product or two, but more often than not the engineering manager talks about product or process technology issues that are understood or cared about only by folks much lower in the organization than those in the staff meeting group.

Perhaps as a leadoff, or perhaps at the conclusion, finance eventually passes out statements and launches into a soliloquy. Variances and ratios are explained and discussed. Typically HR, IT, supply chain, and engineering mentally disengage at this point since little they do affects the short-term fluctuations. They are all on fixed budgets, and so long as they stay within them they are below the radar of anyone worrying about monthly profits.

The financials are usually the subject of a mildly adversarial discussion between sales and operations, with operations suggesting that if sales team members would sell closer to their forecasts all would work much better, and with salespeople suggesting that if operations had things more under control the task of taking operations output to the marketplace would be a whole lot easier.

The task of the boss is often to keep things moving on schedule and, occasionally, to arbitrate any conflicts. Then it ends with little or nothing in the way of solid results. Few action items and few problems are actually resolved and put to rest.

The goal of communications is met—after a fashion, at least. Each functional manager has a chance to let the rest of the department heads know what is going on in each particular area. The only times all staff members are fully engaged tend to be when somewhat mild topics such as the best date for the company picnic arise. The knottier questions of why things are not happening according to plan are usually left after some minor rock throwing back and forth between a few staff members. Those problems will be left for the boss to work out later. Based on the discussion in the meeting, and perhaps the outcome of some departmental performance metrics, it will be up to the boss to assess where the greatest fault lies and to have hard discussions with the manager involved at a later date to try to get things back on track.

At an increasing number of very high-performing companies, however, staff meetings are less frequently held, and the nature of them is radically different. In those companies, rather than having functional managers, a handful of value-stream managers stride in. They each discuss profitability, progress toward breaking through constraints, increases in the value proposition as perceived by customers, and advances toward strategic goals of growing market share.

Each of those managers speaks for all of the functions within the value stream since each leads a fully integrated team that takes customer orders and new products from their inception to their final delivery. Such staff meetings are rarely called for because the ultimate objective of such meetings—communications—is the value-stream managers' stock in trade. They spend all day facilitating cross-functional communications among the value-stream team, assuring everyone is focused on the most critical constraints and on continually enhancing flow to achieve the best results.

These value-stream managers measure themselves and their teams in terms of bottom-line results: profits and growth. They care only about total costs, and they spend little time with detailed internal measurements. Their pricing strategies and capital investments are driven by optimizing capacity to achieve the most economical total outcome, without regard for how it might affect any one particular area of the plant or the business.

The old department heads—the senior managers who once ruled their functional silos—sit off to the side in a much different role. They are advisors, mentors, teachers, and internal consultants. They worry about the long-term and the strategies of the organization. Their senior positions, along with their storehouses of knowledge that took them to those positions, allow them to exert a great deal of influence, but they leave the day-to-day and week-to-week management of the company to the value streams. They work together with the other senior managers as a team to collectively guide the value streams and the company to its long-range objectives.

The purpose of this book is to show the teams of functional managers in the first staff meeting scenario how the best-performing companies climbed out of their silos and into the stream of continuous value creation and growth. The path is not one of having to learn a new set of buzzwords or to master a new technology. It certainly is not a path that requires management to master a Japanese vocabulary. In fact, the path is one that eliminates the clutter and drastically simplifies the business.

Over time we have created a management scheme that calls for too many narrowly focused experts in each of the organization's areas of business and technical expertise, and then have entrenched those experts in functional silos detached from each other. We have tried to link those silos together with increasingly complicated systems and processes—ERP systems that require years of training and experience to master, financial controls that make little sense to anyone outside of the accounting realm, and detailed metrics of performance built on a hope that if each individual or department hits some performance goal then somehow the organization as a whole will prosper.

Communications are indeed a difficult and serious problem because we have created specialist managers who do not live in similar worlds or even speak the same business language. While the sales and marketing people talk about brand strategies and the operations people talk of throughput and pull systems, the others around the table often have little idea what the others are really saying. More important, they often do not care very much since there is little linkage between the departments. It takes much more than an hour or two every week to break down these barriers.

Far from linking the business together, this entangled approach to business has served only to drive wedges between departments and managers who, as they become more experienced in their own areas, more and more lose any connection with the overall business. To run these entangled organizations, we need people with advanced degrees in business—but often little real knowledge of our particular business—who specialize only in trying to organize and keep this disjointed group on track.

The fault perhaps lies more with the consultants, authors, and academic experts and less with management. Many of the ideas that have been classified as fads and buzzwords have served only to further confound when they should have simplified. The very basic concept of identifying the constraints to getting things done has become shrouded in mystique and complexity, requiring something called a "Certified Jonah" to implement. The basic idea of Six Sigma—make sure your processes are capable of doing what you need them to do—calls for certified black belts to explain and put into play. Worst of all, the nuances of the Toyota Production System, renamed Lean Manufacturing, have been widely misrepresented and misunderstood, in no small part because the system seemed to require a crash course in Japanese to decipher the role of *senseis*, or masters. The commonsense idea of a *Kaizen* event requires an outside expert to show management how to get people together to solve problems.

Finance and accounting are little different. The process of accounting for the simple proposition of buying material at a good price, processing it efficiently, and selling it at a profit has become enormously involved. Creating standard costs and measuring minute variances in three or four different categories, rolling the money through generally accepted accounting principles (GAAP) rules that wash spending through inventory and spit it out in a time period other than when it was spent, then cranking out reports far detached from the simple proposition of spending less money this month than last have led to accounting departments filled with people who generate reports few in the company really understand instead of people who spend their days helping the rest of the organization improve real bottom-line results. The chasm between real money (i.e., cash) and financial reports has grown so great that most people cannot see the other side.

Staff departments have sprouted up to accommodate the need for systems experts and masters of the fads. What should be strategies and techniques for getting closer to the heart of things have spawned more complexity. In addition to the regular workload, people must make time for the *Kaizen* and Six Sigma efforts. ERP has become so pervasive that no one really understands it all. A simple change in labor reporting on the shop floor requires meetings and extended projects of people from accounting, supply chain, production, and IT to work through all of the implications of the change.

In the end, although each of these additions to the business has been well intended and each has merit, their accumulated track record of enabling the company to enhance value for any of the shareholders has been dismal. Worse, even when value is enhanced, it is difficult to sustain.

The key to long-term sustained value is to eliminate all of the noise, waste, and clutter and to align the business on the commonsense, fundamental issues that determine success. We must move people away from defining success as mastery of their fine-tuned specialty. All that matters is creating superior value—relative to both customer expectations and to competitor capabilities—for the customers and to turn that customer value into growing profits for the owners. Along the way we will increase the financial security for the employees and suppliers and make a stronger contribution to the communities.

While the path we will map is one of simplification, it is not necessarily an easy one. The staff members responsible only for their own area of specialization will become a team of people, each responsible for a segment of the overall business results. There will be little room to blame other people or other departments for shortcomings. Just because problems call for practical solutions, it is not promised that these solutions will be effortless. What is promised is that everyone will be working together on the problems with the greatest bearing on the company's results.

The job of the leadership group will be to change a team of horses pulling in different directions—and often against itself—into a team where the members are all pulling together in the same direction. The starting point is for the leadership group to begin pulling as one, and then to structure the entire organization to do the same.

We are going to discuss the big picture: what value streams are and how to structure the business in this powerful manner to align everyone with the things that matter the most. We will also present the idea of value enhancement—what it is and how it, rather than cost reduction, should become the driving purpose of management. We will talk about the culture and decision-making principles that can propel the business to much higher levels of performance. The essence of this section is learning to think differently. Engaging everyone in the organization in continuously increasing the rate of flow through the factory to the customer will replace traditional thinking about how cost is optimized.

In the remainder of this section, we will describe some basic management structures we believe are necessary to put this new thinking into play. We will boil constraint management down to its practical terms and lay out a sound approach to accounting that enables everyone to spend money where it adds value and to stop spending money where it doesn't. Quality can be managed and Six Sigma thinking applied without the need for high-cost experts, and we will show how. Performance metrics that keep everyone focused on the bottom line will be outlined. A chapter is devoted to how the supply chain can be structured and managed in a manner aimed at continually driving improved value to your customers. Last, we will talk about human resources management practices. People are at the heart of every business success or failure and how the business relates to the people who drive it is vitally important.

In the second section of the book we will describe how to put all of this into practice. Managing the day-to-day operations and pricing factory capabilities in a manner that takes the best value proposition the company can muster to the marketplace for the greatest possible profits are at the heart of it. We will also discuss an ongoing process of strategic planning that will enable you to move away from annual goal setting and to replace it with a dynamic process of seizing opportunities and engaging the entire company in the effort to make the most of them.

Section I

Thinking Differently

Learning new things is not particularly tough. We do that every day. In fact, in this age of information technology we have to keep learning at a breakneck pace just to survive. The adage about old dogs being unable to learn new tricks is not true, as demonstrated by a great number of "dogs" who are getting gray around the temples running and driving some of the most amazing technology companies. What is tough is unlearning what we have long believed to be true—and that seems to be difficult for dogs of any age.

There was a day when a Swiss watch was the paragon of technical precision. To own one was something of a status symbol, and there was no questioning the accuracy of the time it kept. There was a huge industry in Switzerland to produce watches, and the necessary skills for precisely machining the tiny gears and components to attain timekeeping perfection was not only a source of pride but also a big contributor to the Swiss economy.

In 1962, a Swiss organization called Centre Electronique Horloger came up with the first practical applications of quartz crystal technology for timekeeping. The technology was much more accurate, smaller, and lighterweight and could be made at far lower costs than mechanical watches. In spite of all of the clear advantages, and even though the technology originated in their own country, the Swiss watchmaking industry as a whole rejected quartz crystal and stuck with their traditional ways.

The Japanese held no such bias, and companies like Seiko arose quickly and adopted this new and better way of building watches. Since then the Swiss have staged a comeback using electronic technologies, but the fact remains that over 60% of the Swiss watch companies in existence in 1970 are now gone—victims of their unwillingness to let go of historically tried and true ideas even when faced with overwhelming evidence of a better approach.

This often cited example of a paradigm shift—a change in the fundamental model of how things work—opens up all sorts of psychological debates. How people can cling to old ways to the point of self-destruction when those ways have clearly been replaced with something better is a mystery. You need to figure out the solution to the mystery, however, at least for yourself and your company because the basic theories of management are undergoing an enormous change, and the people who insist on managing in the twenty-first century by what are basically 1960s methods are going the way of the Swiss watchmakers.

The problem with a paradigm shift is that it sets the knowledge base back to zero. The day the company switched from making mechanical watches to quartz crystal, the knowledge and experience an aging Swiss craftsman accumulated over 30 years of watchmaking was useless. A kid fresh out of high school knew as much about making quartz crystal watches as that old pro. No doubt the old guy could have learned the new methods—the old dog is very capable of learning the new trick—but the old dog was unwilling to let go of his old tricks. That is understandable. It requires a great deal of self-confidence to take the attitude that "I learned and mastered the old methods, and now I will learn and master the new methods just as well or better." Many people live in fear of not being able to duplicate their past success. They resist change because they may find themselves at square one again and incapable of making it back to the top.

The challenge an organization faces in transitioning to an entirely new way of thinking and operating is that a lot of old dogs are sitting around the senior management staff meeting, who have attained their positions because they have mastered the old ideas. They have devoted a lot of years and a lot of hard work to being very good at managing their functions the way they were taught, which was not only the best way, but the only way. Now they are being asked to collectively write off the skills that have driven them to the top, and that is a lot to ask.

It is not as bad as all that, however. In fact, much of the success that managers have enjoyed came as a result of their ability to work around—rather than within—the barriers put up by the old methods. People succeed because of their ability to establish good cross-functional relationships. Dismantling the old functional silos and replacing them with value streams only formalizes the way managers have tried to work in the past. It is more an exercise in simplicity, common sense, and doing things the way they know deep down it should have been done all along. When we focus the financial equation on creating value, rather than on rampant cost reduction, we are stating the obvious to most managers who have battled to meet across-the-board cost goals with their best judgment of what should and should not be cut without being unfair or unwise. The culture we are espousing is the culture managers want to create but that has been held back by some outdated notion that managers had to be tough and objective. The culture we will espouse is really no more than saying, "Don't change when you walk through the office door on Monday morning." We are saying managers should be the same person on Monday they were on weekends with their families and at their place of worship. The same combination of knowledge, experience, compassion, and common sense that makes someone a successful spouse, parent, and neighbor will make them a successful manager.

So the changes may be procedural, and some of the reporting relationships may change. However, in the end selling is still about communications, and manufacturing is still about organizing people and machines. Engineering is still about transforming materials, and human resources is still all about the messy endeavor of dealing with people with all of their quirks and idiosyncrasies. The core of your experience and success will still apply.

That said, there is still considerable change involved in taking the company to a much higher level of performance. That is to be expected, of course. If it were easy, everyone would have done it by now. A high degree of confidence in yourself and in each other is necessary, as well as a commitment to the principles that drive such lofty performance.

In the next few chapters we will be discussing ideas. Don't worry—this is not just another theory and philosophy book. We will get to the good part: the meat of the issue and specific action items. But first we have to establish a couple of basic principles. The action item to take from them is primarily in the way you think. The best way to absorb them is to do so as a management team. The members of the management group should read a chapter and then get together to discuss it, to question the implications, and to come to a consensus of what the chapter means to each manager individually and to the organization as a whole. Each manager should do his or her best to personalize the message in the chapter and to avoid falling into the trap of thinking about how everyone else should change.

In later chapters we will get to how these ideas will be put into action in a fully integrated management system, so don't let yourself get caught up in the problems you might see in putting the ideas into practice. The important thing at this stage is to evaluate the ideas on their own merits. Ask yourself whether this is a culture in which you would like to work and one that would improve the overall work ethic, energy level, and environment of the company. Do not get too hung up on all the obstacles you might see to maintaining profitability when you are saddled with some of the restrictions the culture might place on you. We will get to that part.

Likewise with the concept of value or the idea of optimizing flow. How these things interact with your accounting system or problems that constraint management might present with your ERP system are not important at this juncture. We will deal with them later. For now, what matters is whether you understand them and whether they make sense to you.

Take as much time as the management team needs to read, reread, and discuss each chapter comprising the conceptual section of this book. Each contains weighty subjects. When the entire team is comfortable with them, move on to the second section, and we will start to build an organization around a powerful, inclusive, culture designed to create maximum value for customers and to drive costs down through a flow focus in a manner much more effective than anything you have tried in the past.

2

The Mom-and-Pop Theory of Management

Whoever came up with the idea that management is little more than a grand game of sudoku—a couple of numbers are given and you win the prize if you can fill in the rest and have it all fit—missed the boat completely. Most of the management thinking that drives companies to be poor performers and lousy places to work began with that philosophy. The theory seems to be, like that sudoku game, that you are supposed to start with investment and profit goals, to fill in the sales forecasts that make those numbers work, and then to set cost targets necessary to fill in the rest of the blanks so that everything fits up and down. Finally, you task the staff with making the numbers happen by any means necessary, and proclaim yourself to be a good leader. None of it is based on reality: just a lot of numbers that look good on paper, numbers that look good to Wall Street analysts and bankers. Ridiculous ideas like *management by objectives* and *stretch goals* arose from this sort of thinking.

Those ideas along with *professional management theory* found their origins in the business schools where you can count on one thing: no one there knows the first thing about *your* business. Ford, Procter & Gamble, Kraft, Sherwin-Williams, Armstrong, Wahl Clipper, Andersen Windows, Boeing, Buck Knives, Westinghouse, Northrop Grumman, Weyerhauser, Broyhill, Cessna, Deere, Harley-Davidson, Parker Hannifin—these are not just the names of big manufacturing companies. They are the names of men who created companies—and none of them had an M.B.A. Their theory of management was to round up the best people they could find, to make something that was a better value for their customers than anything anyone else was making, and to make it at a good price and as low a cost as possible. Most of them would be appalled at what their namesake companies have become.

Management is not about numbers and computer systems. It is about people. It is all about getting the people working for you and your suppliers to combine their brains and brawn to create things that are helpful to the people who represent your customers. Numbers and ratios are nothing more than a very feeble attempt to describe and quantify the interactions among all of those people.

You already know how to get people organized around a common cause for everyone's benefit. You do it every day in your most important role: being a leader within your family. In just about every aspect, running the business is very much like leading your family and managing your home. In that regard, perhaps the title of this book, *Simple Excellence,* is inaccurate because there is nothing simple about most families. However, it is something you know and execute based on simple principles. Make sure everyone is engaged, involved, and working to his or her potential. Make sure the cash gets managed and the finances are strong balancing short-term and long term-needs. Do your level best to be sure that when your kids leave the nest you are sending the best quality product you and your spouse can turn out into society—young people with their heads screwed on straight who will be happy and successful and will do the family name proud.

You succeed at this family leadership challenge by working hard, by assuring that everyone in the family shares the same goals and believes in the same principles, by communicating these well and often, by making sure everyone knows the plan, and by being smart and disciplined with the family bank account. Mostly you do it with compassion and love—sometimes tough love—but always with an absolute commitment to each other. That also pretty well describes the founders of the great manufacturing businesses mentioned previously: not a lot of complicated management theory.

Management really is simple. Put all of the complex financial books away and focus on cash. Henry Ford said that all he needed to know about accounting was whether there is more money in the bank at the end of the week than there was at the beginning. That is pretty much the same way you gauge your family's financial condition, and it is absolutely correct. No matter what the rest of the financial statements say, if you keep taking in more money than you are spending, things will work out very well.

A whole lot of managers who thought otherwise learned a hard lesson over the last few years when the credit sources went up in smoke. Those who held the notion that cash is just another asset—no more important than inventory and accounts receivable—the ones who believed the DuPont return on investment (ROI) model had anything to do with reality found out the hard way that there is a very real difference between actual money and a "current assets" number on a financial statement.

The volumes written on human behavior and how to motivate, analyze, measure, whip into shape, and otherwise manipulate people into doing what you want them to do should go into the dumpster. Managing people is no more complicated—and no easier—than the Golden Rule. Treat people with the same level of respect, honesty, and fairness you want, and you will be a very good manager. Managers who think they can fool or manipulate anyone are about as successful as husbands, wives, parents, or children who think they can fool their own family: they fool only themselves.

You cannot run the purchasing function in your business any different from how you would at home and expect good results. Sourcing the core product you sell with the lowest bidder makes about as much sense as driving your kid all the way across town to a free clinic, bypassing a better-qualified doctor along the way to save the copay. The same is true of sending your core product to China. It is akin to sending your young children off to boarding school. A manufacturer outsourcing manufacturing and a parent outsourcing parenting so they can both focus on more important matters: what could possibly be more important? Selling? Managing the money? There won't be anything worthwhile to sell or any money to manage when the product representing the very reason for the company's existence gets back to you and is something different from—and inevitably worse than—your expectations. Can you name a single company that was successfully manufacturing in the United States whose fortunes have improved—its products are better and its business is growing—as a result of moving manufacturing to China?

Investment decisions are the same. Pay good money, buy the very best for the things that are important, and look on eBay for the things that are not. The labor cost is a minor consideration. The idea that every machine decision should be based on an analysis of cost reductions from the new machine compared with the old methods using some hurdle rate or discounted cash flow to trigger the go or no-go decision is silly. If that were the right way to make purchasing decisions, no one would ever buy a new car.

It could never be cost justified. When the old one gives out you should buy the cheapest wreck you can find and drive it into the ground. You don't do that because things like reliability and safety enter into the decision as well as quality of life. Just because the originators of ROI theory did not know how to put a value on such things does not mean they are any less relevant to the business decision than they are to your personal decisions.

The books on organizational theory should find their way to the same scrap heap. All of them are based on some idea or another of command and control, limiting communications and limiting the involvement of people in decision making. What form of human endeavor is bettered by restricting communications and narrowing the input to decision making? None that we can think of. In fact, the more people who know what is going on and provide more points of view to consider, the better the decisions that result and the stronger and healthier manner in which problems are resolved. A hierarchical organizational structure, intended to enable the person at the top to bark, "Jump!" and have that order flow like the wave at a football game into synchronized jumping throughout the organization is not only wholly unnecessary but also ineffective. The best organization is, again, like the family. Everyone should know what is going on up, down, across, and diagonally as much as possible. And everyone with an idea should feel free—should be urged, in fact—to contribute to every problem and decision. An autocratic chief executive officer (CEO) or vice president is about as effective as an autocratic spouse or parent.

There was a placard on Albert Einstein's wall that said, "Not everything that can be counted counts, and not everything that counts can be counted." This from the guy who went down in history as the most counting, mathematically oriented genius ever. Everyone knows this is true except, of course, the business theorists who suggest that everything important must be measured and that things that cannot be measured are inherently unimportant. That makes about as much sense as measuring yourself as a parent on your kid's grade point average, how many times he made his bed last week, and his little league batting average because you know how to boil them down to numbers while writing off his levels of courtesy, responsibility, and kindness to others as insignificant because you don't know how to put a number on them. We would all take a courteous, responsible kid over one who can hit a baseball thrown by another 10-year-old any day.

Somehow we have propelled management to a level at which common sense and the same basic values that serve us well in the rest of life do not apply. We have bought into the Godfather principle that separates "business" from "personal." We have come to believe that the things we put into play with our families and strive to strengthen in places of worship on the weekend have no place in the office from Monday through Friday. It is high time to take a big step back and take a hard look at the whole business of management.

The early histories of Ford, Procter & Gamble, Kraft, Sherwin-Williams, Armstrong, Wahl Clipper, Andersen Windows, Boeing, Buck Knives, Westinghouse, Northrop Grumman, Weyerhauser, Broyhill, Cessna, Deere, Harley-Davidson, and Parker Hannifin are stories of hard work, common sense, and absolute commitment to making the best products possible. The same is true of just about every business, regardless of where it is or how much it may have grown over the years. The machine shop on the corner began because someone had a conviction that he could make something better than the other sources available to some segment of some market. It succeeded because good people worked very hard and proved that point. The best companies we have had a chance to observe— especially those that have succeeded and even grown in the face of the global recession—are the ones driven by those same simple principles: simple excellence. They do not get tangled up in far-out management theories, and they do not spend a lot of time reading and pondering the latest issue of the *Harvard Business Review.*

The common trait among these best companies is their ability to concentrate on the fundamentals and to keep things simple. In no particular order, people, cash, their product, and their customers are the priorities. Everything else is secondary. Common sense and core values—a very keen sense of right and wrong—trumps management theory every time. The more complicated a new idea is, the less likely the companies are to adopt it. The most striking difference between the excellent companies and the not-so-excellent ones, and the area in which they are most like watching family decision making in action, is in the time frame of their decision making. The future is always more important than the present. Even though companies like Wahl Clipper are enormously successful, they operate as though the future will hold formidable challenges and they have to focus on working hard to do the right things—the difficult things—today if they have any hope of meeting those challenges. It is the same as the family in which everything is firing on all cylinders but father, mother, son, and daughter still get up every morning and head off to work and school and then come home and do their chores and their homework. Today is merely preparation for tomorrow—and tomorrow never gets here. It is the polar opposite of the firms obsessed with the current quarter and what Wall Street wants now.

It is ironic that most companies start out very focused and lean. It is only through success that they become bureaucratic and wasteful. Then they compound the bureaucracy and waste with management theory to help them wade through the unnecessary complexity they have created. Companies want their managers to be more entrepreneurial, but they expect that entrepreneurial spirit to thrive within the confines of reports, meetings, metrics, structures, and financial ratios that no entrepreneurs would ever waste a dime or a minute of their time implementing.

The simplest, leanest, and best-managed a company will ever be is in its infancy when senior managers wear a lot of hats, everyone is fully engaged, and the business operates with a sense of camaraderie forged by a shared sense of urgency. Cash is king in the start-up, and spending is tightly controlled. The start-up hangs on every word spoken by the customer, and suppliers are chosen cautiously. It takes success to destroy all of that. The road to excellence requires undoing the fruits of success and returning the company to its entrepreneurial roots and focusing simply on the few things that matter.

3

It's All about Value

Go to Google and enter the search terms "cost behavior regression analysis," and you will be either shocked or awed by the 373,000 results, most of them on Web sites that end in ".edu." An enormous amount of intellectual ability is still being wasted on sophisticated models of manufacturing costs and how to predict and control them. Instead of harnessing more computer power to look at costs in more complex manners, how about this instead: spend more money on things the customers will pay for and less money on things they won't pay for. It really should be no more complicated than that. Most companies, however, have their cost management so tangled up they have no idea what adds value and what doesn't.

While the mathematics of success are quite simple, they are a bit more complicated than "sales minus costs equals profit." That equation is certainly true from an algebraic standpoint; however, it is a dangerous proposition with which to manage. It leads to thinking that prices—one of the two variables in the sales part of the equation—have something to do with costs. They don't, but, again, that is a matter of common sense. Prices are a function of the value the product offers to the customer. All managers know that, but most fail to make the connection.

When you are strolling the aisles at Walmart, do you care what the manufacturer spent to make things in deciding what to buy? Of course not. You care about the price compared with what the product can do for you. And when you see two similar products you are not about to pay more for one than the other because the manufacturer of the more expensive one happened to waste more money on paperwork or material handling than the other. You go through a mental exercise in which you are assessing three things: what you think of the quality of the two products; how well it fits your intended use of the product; and how reliable you think the two products will be. Somehow or another you add up those three things and put some sort of subconscious value on them and compare that value to the price. You don't always buy the cheapest product. In fact we rarely buy the cheapest product; if we did there would be only one type and brand of everything on the shelf—the cheapest one. No, you buy the one with the best relationship of that value you assigned to the price charged. We commonly refer to this mental math as "getting our money's worth."

Your business is no different. You do not succeed by being the low-cost producer. You succeed by being the best-value producer, and value is a function of three variables: quality, utility, and reliability. If the ratio of the price you are charging to the value you are providing is better than the other guy's, you will more often than not get the sale. If that value proposition is not as good, you lose. This should be intuitive, but it clearly is lost on many manufacturers, hence the exodus to China. Driven by the destructive math of sales minus cost equals profit, they believe that if only they can get the cost down they will be more profitable. If they were not profitable to begin with, the problem was the value proposition, not the cost, and the notion that they can get their product made cheaply in China or anywhere else without affecting the value is ludicrous.

The major brands are under siege almost across the board for the simple reason that they have missed this point. It is estimated that by 2012 over half of the retail items purchased in the United States will be privately labeled or store brands—and why not? If the big brands are going to bail out of manufacturing and buy lesser-value goods from China, why shouldn't Walmart and Kroger cut out the middle men that add no value and buy the same-value products directly from China themselves? And, as consumers, why wouldn't we want the retailers to do this for us?

Walmart takes a beating for hammering American manufacturers on price—driving them to China, so the wailing goes. All it is doing is looking out for its customers' best interests and assuring that its customers get the best value for their money. Consider the following two tales.

The Wahl Clipper corporation,[1] makers of hair clippers for both the professional and consumer markets, has taken over the shelves at Walmart,

---

[1] All references to Wahl Clipper Corporation are based on the personal knowledge and experience of author Bill Waddell from 2005 through the present as a consultant to Wahl Clipper, and his work as VP of Global Operations, as well as information presented by Wahl Clipper personnel at the Lean Accounting Summits in 2007 and 2009 in Orlando, Florida, and in 2008 in Las Vegas, Nevada.

with higher-cost and higher-priced products, driving their 100% Made in China competitors off the shelves. That hardly squares with the charge that Walmart cares only about price and is "driving manufacturers to China." When it comes to hair clippers, Walmart has been quite happy to do the opposite. In fact, Wahl has grown market share by huge bounds with products priced 30% or more higher than other companies' by offering hair clippers that are perhaps twice the value. Consumers don't want cheap—they want their money's worth—and they believe they get more for their money when they pay more for the Wahl product.

A story in the *Wall Street Journal*[2] recently described a young lady who had been buying a $17 bottle of Procter & Gamble (P&G) brand shampoo. Times being tough, she switched to a $5 bottle of P&G brand shampoo and learned that she could not discern any difference. The price gap was not based on value but on an image created through brand management strategies. Having learned that she had been duped all along by slick advertising, she made it quite clear that, no matter how good things are in her economic future, the odds of her ever falling for the advertising con that created the false illusion of value are slim indeed.

The companies that outsource manufacturing in pursuit of low cost are, by and large, racing each other to the opening price points in their categories. In Wahl's case, all of their Made in China competitors are fighting each other for the bottom shelf at Walmart where the $14.95 hair clippers are sold, while Wahl happily takes the rest of the shelf space where 80% of the volume is. The competitors got their costs down, to be sure, but lost the business in the process. And P&G proved, once again, the wisdom in the Abe Lincoln adage about it being impossible to fool all the people all the time.

The problem with cost reduction is that some costs add to the value of the product in the eyes of the customer while others add nothing at all. It is ironic that the companies pursuing an outsourcing strategy are typically looking to reduce the value-adding costs (i.e., direct labor and materials) and are actually adding to the costs that do not add value (i.e., overhead and administrative costs). It is wishful thinking to believe that the value-creating costs can be slashed by huge orders of magnitude without affecting the value they create. This is the folly of the sudoku school of management.

---

[2] http://online.wsj.com/article/SB124946926161107433.html; see also Byron, E, 'Tide Turns Basic After the Slump', *Wall Street Journal*, August 6, 2009.

Managers who see manufacturing as a grand exercise in numbers rather than as a shop-floor business where material is transformed into something tangible fall for the idea that anybody can make anything just as well as anyone else. In short, they see manufacturing as a commodity, when it is anything but that. They think that the product they can get from China for $5 is the same as the product that comes from their own shop floor for twice that amount simply because both their people and the Chinese are working to the same set of engineering specifications.

The dynamics of excellent manufacturing are invisible to the professional manager who runs things by theory and numbers, but they were very real to the original entrepreneur who started the company. In companies such as Wahl, the specs are the starting point—not the end—and engineering and production are continually working together to make miniscule corrections and improvements that constantly make the product a little better. Defects and customer input are continually and quickly fed to the shop floor for course corrections. The opposite takes place when the product is made by a cheap labor force on the other side of the world. Chinese workers with no personal knowledge of the product because they couldn't afford it if they wanted it are neither empowered nor able to work with engineers and customers from the other side of the globe to fix or improve anything. Instead, it is an ongoing, losing battle to produce to the specifications and to try to filter out all of the defects being produced. Wahl clippers from Illinois keep getting a little bit better, while the quality of the competition's clippers stays the same or even degrades. So Wahl laughs all the way to the bank, while the low-cost manufacturers cry all the way to bankruptcy court to file their restructuring plans.

This doesn't mean that reducing costs is not important. It is very important. Knowing that all costs are not the same, however, is the trick. High-value producers cut the waste ruthlessly but are very careful about cost reductions to activities that create value. Their goal is not to be the low-cost producer; it is to be the best total-cost producer—that is, with the greatest percentage of spending going to value-adding activities. If 60% of your company's total spending is going to the quality, utility, and reliability of the product while the other company devotes only 50% of its expenses to such activities, yours gets to charge higher prices and grow market share, and it doesn't. So the objective is to continually improve the composition of total spending rather than to simply reduce all costs. If along the way you can reduce the total costs in addition to improving the percentage devoted to value creation, so much the better.

While most accountants and most companies worry about relatively meaningless breakdowns of spending by "fixed versus variable" and "direct versus overhead" or about spending in functional departments compared with an annual budget, the excellent companies worry about value adding versus non-value adding. Whether a cost is contributing to premium pricing and greater customer value is a whole lot more important than how that cost tracks on a regression analysis chart. Who spent the money is not nearly so important as why they spent it.

The challenge most companies face in determining which costs add value and which don't is that they don't really know how the customers are going through that "getting their money's worth" mental exercise. They don't know that because they quite often don't really talk to the actual customer who makes the decisions. Walmart places no value on hair clippers at all. They leave that up to *their* customers. Walmart's objective is to provide the products their customers believe have superior value. So in Wahl's case, the definition of value is provided by end consumers—not the Walmart buyer. To gain that knowledge, a lot of people are walking around northern Illinois with bad haircuts as Wahl is eagerly putting new clippers into people's hands to see how they like them—fielding complaints from irate parents from time to time when their kids signed up for a free haircut from a local amateur with a new Wahl product.

Packaging machine builder Barry-Wehmiller[3] tries to get the customer face to face with its employees who are actually building the machine to be absolutely sure it provides everything the customer wants and needs. It knows that not everything can be boiled down to a specifications sheet. The buyers in the customer's business are rarely the people actually using the product; to them it is simply a numbers exercise. Barry-Wehmiller is trying to get past the middle man and talk to the people who really assign value in all of its terms, putting them in direct contact with its people who will create that value.

---

[3] See http://www.barry-wehmiller.com. The Barry-Wehmiller history and philosophy are well documented throughout the Web site; additional information on Barry-Wehmiller from Bob Chapman keynote address at 2008 Lean Accounting Summit in Las Vegas, Nevada, author Bill Waddell conversations with Bob Chapman in 2008 and 2009, and from company literature and video provided by Barry-Wehmiller to author Bill Waddell of internal speeches made by Bob Chapman at Barry-Wehmiller on unknown dates.

At the BAE Systems plant in South Dakota,[4] production folks who took suppliers' components out of the box and used them to build missile systems were brought face to face with the last production people at the suppliers' plants—flipping things around but trying to assure that the suppliers knew exactly how BAE Systems determined value. Beyond parts specifications the actual users of the components were able to tell the actual producers about the best-parts orientation within the box, part-labeling issues, box sizes, the best quantities per box, and the best size of the box—lots of things that typically don't find their way onto purchasing specifications but that make the difference between satisfied customers and indifferent ones.

It requires deep knowledge of the product and its life after it leaves your plant to truly understand how that "getting their money's worth" thinking plays out. This is the real downfall of the "professional manager" school of thought. Going back to those original names—Ford, Procter & Gamble, Kraft, Sherwin-Williams, Armstrong, Wahl Clipper, Andersen Windows, Boeing, Buck Knives, Westinghouse, Northrop Grumman, Weyerhauser, Broyhill, Cessna, Deere, Harley-Davidson, and Parker Hannifin—they were the furthest from professional managers. Rather, they were guys who knew their product inside and out, and they would have scoffed at the notion that anyone could run their namesake companies because they had an M.B.A. and, therefore, could run any business. On the other hand, Thomas Barry and Alfred Wehmiller—a couple of old St. Louis machine builders—would be very proud of the way Bob Chapman is running the company they created way back in 1885. And Leo Wahl would be pleased to know that his grandson has an engineering degree and runs the company from the position of knowing how clippers work as well as anyone there.

Only when management knows the products, how they are made, and exactly how customers use them can they determine which expenses add value. Lou Pritchett, the sage P&G leader from before they became a brand management company and were focused on products and customers, said, "The primary job of sales is to represent the customer to the company, not the other way around."[5] The sales and marketing function has to be fully engaged in the critical task of reaching through and around the

---

[4] From work author Bill Waddell did with United Defense Corp. (subsequently sold to and now part of BAE Systems) in Aberdeen, South Dakota.

[5] http://www.loupritchett.com/htm/management.htm from 'Lou Pritchett on Management and People.'

buyer's office and getting close to true customers—then making sure the entire company knows exactly what those people need and want.

The most important line on the financial statements should be a very clear and bright one that indicates the expenses above the line that add value and those below the line that do not. This breakdown has to be very clearly understood by everyone in the company—if the expense adds value then the activities behind that expense have to be executed perfectly and continually improved. The expenses below the line have to be continually reduced and hopefully eliminated at some point.

Note that just because an expense might be necessary—like paying auditors or the cost of complying with the Occupational Safety and Health Administration (OSHA)—doesn't mean the costs are value adding; it means only that they can't be eliminated entirely. However, they certainly can be minimized. It is also important to note that the segregation between value adding and non-value adding applies to spending and activities— not people. Just because someone may be assigned to a task that does not add value does not in any way mean that the person has no value. A multibillion dollar manufacturer[6] (that shall remain nameless to avoid embarrassing them) set off to define value adding versus non-value adding and decided to call all payroll expenses "value adding" to avoid hurting employees' feelings by insinuating that they are not valuable to the company. This is an admirable intention, to be sure, but one that completely defeated the purpose of the exercise.

The bottom line, however, is that understanding the value of what you make is fundamental to success and that the most important role of sales and marketing people is their continual enhancement of the company's understanding of value in the eyes of the customer. There is nothing for them to sell or market if the company is not creating value.

---

[6] Comment made to author Bill Waddell at 2009 Lean Accounting Summit in Orlando, Florida by the CFO of a Fortune 500 company.

4

The High Cost of Poor
Cost Accounting

If your company uses standard costs, the only thing that is certain is that you don't know what your products cost. It would be great if the allocated, fully burdened, standard cost were actually a reflection of the cost to make something, but it is actually nothing more than the product of some fairly simple arithmetic with no significance whatsoever.

To understand just how little such costs have to do with reality, you have only to calculate the fully burdened cost of reading this book. To follow the same logic as accounting you have to include the wear and tear on the book. Let's say you paid $39.95 for it and someone thinks that it can be read 100 times before it starts to unravel a bit—so we'll call it 40¢ for the book itself. Then of course we have to figure in the six hours you spend reading it at your hourly rate plus the cost of your payroll associated costs, including the taxes, something to accrue your vacation time, and also an allocated portion of your medical benefits. The fact that your vacation time and monthly medical insurance premiums will not change one iota whether you read the book has no bearing on the subject—just like when the standard costs of your products are calculated. So we are going to put you down for a fully loaded rate of $75 an hour times the six hours for a labor cost of $450—plus the 40¢ for the book. Then we have to assign the cost of the floor space in your office, the light and heat bill, depreciation on the chair you are using—let's say about $25 an hour times six, for another $150. So your company is out $600 plus the 40¢ for each person in the company who reads this book.

It makes you think twice about having all of your staff—all eight of them—read it. That will result in a $5,000 hit to profits this month, right?

Wrong, of course. If you have your staff read the book—eight copies plus one for you is nine—at $39.95 a book, it will cost you about $360, and anyone in accounting who wants to accrue a $5,000 expense for the book should have his or her head examined.

It is simple, just-like-you-do-it-at-home common sense again. The food budget at your house is very unlikely to include accrued depreciation on the oven, the pots and pans, and the dining room table. Such numbers would be a waste of time to calculate and as meaningless as $5,000 for book reading. Why would anyone think that applying this same logic to the cost of the products you make would result in numbers any more meaningful?

The reason for calculating such numbers is that the government says you must to comply with generally accepted accounting principles (GAAP) when you pay your taxes. If you have to deal with commercial lending officers at the bank, they probably want GAAP statements, too. This, however, is hardly a reason for using such data to run the business.

Consider a case in which your newly married children come to you and say, "Mom, Dad, we are seriously thinking about having a baby, but we aren't sure we can afford it. How can we figure out what it will cost us?" Would your answer be to advise them to read page 44 of the IRS Form 1040 Instructions where they will find the Child Tax Credit Worksheet? Or would you sit down with them and go over their family budget and help them determine the real changes to their monthly cash flow—up and down—that are likely to result from having a child? That, of course, is a rhetorical question. No one would use government figures to make real decisions for anything that is not directly related to taxes. You would not haul out last year's tax return to decide if you can afford a vacation this year. You would get out your bank book and go over the family budget with a sharp pencil.

Thinking the complicated calculations resulting in standard costs are any more relevant is silly. Those numbers are good for filling out forms for the tax authorities and nothing else. They certainly should not be used to make any important business decisions—like pricing or whether to outsource anything. If you want to decide how much it will cost to have the staff read this book, get rid of all the allocations and accruals—those are just accounting terms for numbers that are not real—and use the only thing that is real: the $39.95 a copy it will actually cost you. The calculated $600.40 a copy is useless, as is the standard cost for your products. After the tax returns are filed, those standard costs should never see the light of day.

The destructive impact of traditional cost accounting—with all its allo-cations, accruals, and assumptions—was first exposed in a big way by Tom Johnson in 1991 in *Relevance Lost.*[7] Since then the issue of manu-facturing accounting has been at the center of hot discussions, to which accounting has generally contributed very little. This is finally changing, however, with the advent of Lean accounting. Jean Cunningham's book on the subject, *Real Numbers,*[8] is very appropriately titled.

The fact is that the traditional accounting practices actually discour-age good manufacturing practices. With roots in the same ROI formulas that say cash is no more valuable than purely paper assets, full-absorption accounting encourages building inventory to absorb more of the fixed overhead and to keep it off the profit and loss statements—and of course discourages inventory reductions for the same reasons. That alone should be proof of the serious errors in standard cost accounting. Anyone with the least bit of manufacturing knowledge knows that building inventories and lengthening cycle times is bad practice in every dimension. It drives up indirect costs, gobbles up floor space, undermines quality control, and cre-ates the need for big complicated computer systems to keep track of it all. An accounting system that says such an inventory build-up is good regard-less of the obvious harm it does is clearly very flawed.

There is an often-repeated joke about a guy whose wife asks him to stop at the meat store and bring home a ham. He does so, but his wife gets after him for failing to instruct the butcher to cut off the end of the ham first.

"Why," the guy asks, "do you always cut off the end of the ham?"

"I don't know," the wife answers. "It just makes the ham better. Besides, that is the way my mother taught me, and her hams were always perfect."

Dissatisfied, the guy calls his mother-in-law and asks her the same ques-tion. He gets about the same answer: it just works better that way, and *her* mother always did it that way. Now his curiosity is thoroughly roused, so he calls the grandmother at the retirement home and asks her why just the end of the ham has to be cut off.

"I don't know why anyone else would want to do it," the old lady answers. "I cut it off because I never had a pan big enough to hold the whole ham."

[7] Johnson, H.T. and Kaplan, R. S. *Relevance Lost; The Rise and Fall of Management Accounting*; Harvard Business School Press, Boston, 1987.

[8] Cunningham, J. Fiume, O. and Adams, E. *Real Numbers: Management Accounting In a Lean Organization.* Managing Times Press; Durham, N.C., 2003.

Members of senior management should ask why as many times as it takes for accounting to give them an answer as to why accounting has to be done this way and who says this is the only way or even the best way to account for manufacturing. In all likelihood, the accounting department will come up empty and have no explanation for the origins and the basis for such absurd assumptions about the financial implications of manufacturing. The few steeped in accounting history will be able to explain that most of it goes back to a hopeless alcoholic by the name of Donaldson Brown who worked for General Motors (GM) in the 1920s.[9] It is done this way because General Motors did it that way and made a lot of money back in the day. Most of it codified into law through an Accounting Research Bulletin formalized in 1953. So you are locked in a battle for the very survival of your company against the fiercest of global competitors armed with financial thinking that has not progressed one iota since the Dwight Eisenhower administration. While every other function of the business is pedaling as fast as it can to keep up with the state of the art, often using technology developed after Bill Gates retired, an accounting function stubbornly insisting on standing pat with practices codified the year Bill Gates was born is unacceptable.

Concepts such as *product* or *customer profitability* have to be kicked to the curb. There is no such thing. Only businesses can be profitable—or not. Every product and every customer contribute to your profitability to varying degrees. The only question is how much of your operating costs they cover. If one of your staff goes out and saves you $500 as a result of reading this book, common sense would say that you got a pretty good return on a $39.95 book. Accounting would say don't let anyone else read the book since you lost $100.40 on the deal. Accounting reports that say you lose money on any particular product are based on precisely the same illogical logic.

Equally pointless is worrying about whether costs are fixed or variable. The answer is yes they are fixed, and yes they are variable. In the short-term

---

[9] Waddell, W.H. and Bodek, N. *Rebirth of American Industry*, PCS Press, Portland, Oregon, 2005. In *Rebirth of American Industry* the evolution of the General Motors management system and the roles played by Alfred Sloan and Donaldson Brown are described, as well as the evolution of Henry Ford's manufacturing and management system and its evolution to the Toyota Production System and the roles played by Taiichi Ohno, Shigeo Shingo and others at Toyota. The origins of the DuPont ROI model are also described as central to the GM management system. All references to these companies and individuals arise from this book and the reader is advised to refer to that book for further information on any of these people or subjects.

all costs are fixed. In the next hour there is next to nothing you can do to change the costs in the factory. They are past the point of no return and are pretty solidly fixed. Over the course of the next five years, everything can change, including where the factory is located and whether anyone in senior management still has a job. They are all variable. So everything is fixed in the short-term and variable in the long haul, and some of both anywhere in between. But so what? Does knowing whether they are fixed by some mathematical definition matter in any practical situation? Even if your clever teenage son provided you with a regression analysis of your electric bill demonstrating that the bill is statistically fixed, you would still tell him to turn off the lights, the TV, and the stereo when he leaves the room.

Whether someone in accounting determines that something is fixed or it is variable, you still want the same thing. You want the organization to find ways to spend less of it—especially if it is something that creates no value for the customers.

The central issue is that you have to get the fog and complexity out of your accounting processes and start to manage the business with real numbers— real, practical numbers put together largely the same way you put together your household budget. In some companies the answer to the questions of what was spent last month and was it more or less than the month before cannot be discerned from the financial statements. They are such a compilation of accruals, allocations, and GAAP-based nonsense that such fundamental information is unknown and, often, unknowable. That has to change, and your accounting folks need to be out in front leading the way—not coming along complaining, kicking, and fighting every step of the way.

Often, before the accounting team members can begin to generate useful numbers their role and responsibility needs to be redefined. In far too many companies accounting employees have taken their audit and fiduciary control obligations to such an extreme that they no longer feel responsible for helping to lead the company. They are allowed to sit on the sidelines as some sort of uninvolved, impartial judges of the rest of management rather than to operate as fully responsible and equally committed members of the management team. That aloof, irresponsible role is what the outside auditors are for. Accounting team members should in no way be tasked with conjuring up numbers to make things look better than, or even different from, reality. They simply have to get out ahead of the lean-accounting, real-numbers curve and lead the company instead of fighting change for no good reason.

The first accountant who raises Sarbanes–Oxley as justification for providing management with useless GAAP-based data to run the company should be told to re-read the *Lean Accounting* book, Sarbanes–Oxley, or both. That is a sure sign of an accountant who (1) doesn't want to help, (2) doesn't really understand Sarbanes–Oxley, or (3) both. In any event, the company needs financial leadership—not 1953 thinking and the work product of a General Motors engineer from the Roaring Twenties—if it wants to join the ranks of the excellent companies. Big companies like Boeing and Parker Hannifin, mid-sized companies like Wahl and Barry-Wehmiller, and companies as small as 40-employee BRAMS,[10] as well as hundreds of others, have tossed GAAP and standard costing out the window and are managing the business with those real numbers of which Jean Cunningham wrote. The rest of manufacturing has to do the same.

---

[10] http://www.leanaccountingsummit.com/ The list of companies pursuing and practicing Lean accounting is continually changing and being updated on this Web site. The companies cited are from this list as well as author Bill Waddell's personal consulting experience.

5

Can We Find a Few Righteous People?

We may not be living in Sodom or Gomorrah, but it sure feels that way to read much of the business news. The folks at General Electric (GE) admit to fraudulent dealings with the U.S. Securities and Exchange Commission (SEC), Chiquita admits to funding terrorism in Colombia, the head of Rio Tinto in China sits in prison for bribery, and the boss at KIA is elated to have his embezzlement prison sentence reduced to house arrest—all that is before we get to Bernie Madoff, Allen Stanford, and the rest of the financial sector where the serious crimes have been committed. Most of the shady dealing that seems to have become common business practice does not rise to the level at which executives go to jail, but greed sure seems to be the order of the day, even at the companies doing well and enjoying sterling reputations.

Smarter people than us will have to figure out how things got to where they are, but one thing we know for sure is that the driving principles of the excellent companies are far from this "every man for himself by any means possible" thinking. Perhaps the best example is Bob Chapman at Barry-Wehmiller. He is living proof that driving the business by a strong values is not incompatible with making money but is actually the key to business success. The packaging equipment manufacturer based in St. Louis has grown its top line by a factor of four since Chapman took over a few years ago, and the bottom line is growing correspondingly.

Chapman sounds more like a Baptist preacher than a business guru. The Barry-Wehmiller Companies "achieve principled results on purpose," he says, "by continuing to develop a growing business sustained through the power of inspiring people toward a fulfilling experience." Larry Culp at the Danaher Group[11] says they succeed because they "retain, attract, and develop the best talent available," and they do so by demonstrating "high integrity" and the "utmost respect for people." Contrary to the view of people as headcount to be minimized in pursuit of maximizing shareholder value—and maximizing the personal bank accounts of senior management along the way—these guys make it pretty clear that something bigger than themselves is driving their business decisions.

We seem to have arrived at a cultural fork in the road, and members of senior management have to decide which path they will take and then must demonstrate their commitment to those principles. Platitudes don't cut it anymore. Companies like LEGO are all too commonplace.[12] It put flowery nonsense up on the walls at corporate headquarters and in the annual reports about people being "indispensable for sustainable business success and for promoting the LEGO Group's corporate values and culture" and then laid off 5,300 of those "indispensable" people so essential to their "corporate values and culture" when they outsourced manufacturing to just about any and every low-cost country they could find. Talking the talk is easy—walking it is another matter entirely. What guys like Chapman and Culp say is meaningful only because their actions demonstrate their convictions.

Greg Wahl says that his heart has to have as much weight in the decisions at Wahl Clipper as his brain: do the right thing by all of the people who depend on the company. So Wahl hasn't laid anyone off in some 35 years—long before the current management team was in place. SC Johnson hasn't laid anyone off in its 126-year history,[13] and it never will as long as someone named Johnson is running the company. Google operates by the straightforward principle, "Don't be evil," and is willing to walk away from the huge market in China to follow that principle.[14]

---

[11] From author Bill Waddell's conversations between 2005 and 2008 with Mark Deluzio, former Vice President and Corporate Officer at Danaher, now President of Lean Horizons Consulting.

[12] LEGO's initial strategy to outsource manufacturing and lay off staff can be read about at http://money.cnn.com/magazines/fortune/fortune_archive/2006/06/12/8379252/ Their subsequent acknowledgment of the failure of that strategy can be read about at http://ing.dk/artikel/96451-lego-her-er-opskriften-paa-succes While the second article is written in Danish, it generally states that Lego now believes that insourcing – bringing manufacturing back from the third parties in low labor cost countries, and building on intellectual capital is the key to success.

[13] http://www.scjohnson.com/en/company/overview.aspx The SC Johnson Web site contains a wealth of information on the company's history, cultures and values.

[14] http://www.google.com/intl/en/corporate/tenthings.html; Google; 'Our Philosophy – Ten Things We Know to be True' Google's dispute with and exit from China were well documented in the media throughout the Spring of 2010.

You don't have to be a follower of any religious faith to be a good manager. You do, however, have to believe that you have a heavy obligation to all of the stakeholders in the business that is at least as great as the obligation you have to the stockholders and as important as the size of your annual bonus check. The Pope stated it well in his recent Encyclical[15] when he wrote, "Business management cannot concern itself only with the interests of the proprietors, but must also assume responsibility for all the other stakeholders who contribute to the life of the business: the workers, the clients, the suppliers of various elements of production, the community of reference."

Chapman agrees. He says that the great obligation and the great opportunity of members of senior management are their unique position in which they can affect hundreds, thousands, maybe even millions of lives. The measure of management success is not financial but is in what they do with that chance to make so many lives better.

Lest any of this sound like a call to social work, make no mistake about it. All of these companies make a lot of money, and their investors do very well by them. That treating people well and taking good care of customers, suppliers, and the community yields great results should not come as much of a surprise. It is a very straightforward, simple, and commonsense proposition. What is complicated are the convoluted rationalizations management theorists conjure up to explain how you can obtain profits while abusing those stakeholders. That takes some very strained logic to explain.

Suppliers take better care of customers who demonstrate loyalty to them. It's not rocket science. The same is true of the community. The local politicians and city fathers and mothers jump through hoops to take care of the big employer that sustains and supports the community. They don't look at the employer that sees their husbands, wives, sons, and daughters as "headcount" quite the same. Again, this is patently simple and obvious, and it takes only a moment to step back and realize it. What is complicated is convincing yourself that the opposite is true.

Old Alfred Sloan, the management "genius" who planted the seeds for what General Motors has become—a failed, government-owned dinosaur

---

[15] Encyclical Letter *CARITAS IN VERITATE of the Supreme Pontiff Benedict XVI to the Bishops Priests and Deacons Men and Women Religious the Lay Faithful and All People of Good Will on Integral Human Development in Charity and Truth;* Given in Rome, at Saint Peter's, on 29 June, 2009. The document can be found at http://www.vatican.va/holy_father/benedict_xvi/encyclicals/documents/hf_ben-xvi_enc_20090629_caritas-in-veritate_en.html

and the epitome of bad management from top to bottom—once wrote that workers simply had to "accept the hazard of the business cycle." In other words, getting laid off when it is convenient is just the way it is, and people working for him had to learn to live with it. They had to learn to save for the unavoidable rainy days, and to think otherwise was a "panacea" and "wishful thinking."

Chapman, Culp, and Wahl don't think that the employees of their companies have to accept the hazards of the business cycle. They know they are critical to the long-term success of the business. Fisk Johnson from SC Johnson believes that "the goodwill of the people is the only enduring thing in any business. It is the sole substance. The rest is shadow." These guys are endowed with enough common sense to know that the people they take such good care of will reciprocate and go above and beyond the call of duty to make sure the company succeeds. The history of the United Auto Workers (UAW)—the folks Sloan and the modern business theorists view as headcount who should learn to live with whatever management dishes out in pursuit of short-term profits—is hardly one of going above and beyond the call. That should come as no big surprise to anyone.

So what does all of this mean to your business? It means senior management as a group has to decide what the business is all about. You have to decide what you are trying to accomplish and what your operating principles are: what are the rules and limits, and how are you going to go about doing business? You have to do a lot better than set the goal as *maximize shareholder value*. That is about as meaningless and uninspiring a goal as can be imagined. Try something like, *Provide superior value to our customers to maximize long-term value for the shareholders and to simultaneously provide security and value to our employees, suppliers, and the community in which we operate.* However you choose to phrase it, defining your goals simply and with common sense and common decency is an essential launching point for joining the excellent companies.

A good approach, once again, is to tie it back to your family. One of the authors was once chastising a young Chinese manufacturing manager for the conditions of the workplace in a factory area. The Chinese manager, with no frame of reference for understanding American expectations for cleanliness, lighting, and safety, asked for some specific direction. The American knew the manager had a teenaged sister and offered that up as the standard: "Ask yourself if you would want your sister to work here. If the answer is no, then you have work to do. When you get things cleaned up to the point that the answer is yes, then it is clean and safe enough." At the next visit to the Chinese factory, the improvement was extraordinary. The "little sister standard" is universal and can be the guide for cultural transformation. Decide what you want the stated goal to be of the company for which your little sister—or spouse, son, or daughter—is going to work. That should be the same goal you set for the people working for you who have bet the livelihood of their families on your integrity and ability.

6

People

"We believe that business enterprise has the opportunity to become the most powerful positive influence in our society by providing a cultural environment in which people can realize their gifts, apply and develop their talents, and feel a genuine sense of fulfillment for their contributions in pursuit of a common inspirational visions." So says Bob Chapman, chief executive officer (CEO) and chair of the enormously successful Barry-Wehmiller Companies. Chapman hammers on themes like his vision of "achieving principled results on purpose" and defining his management team as L³—the Living Legacy of Leadership—bringing together the tools of Lean manufacturing with Barry-Wehmiller's unique people-centric beliefs.

While some of the things Chapman says and writes sound hokey, under his leadership the Barry-Wehmiller Companies have experienced amazing growth and financial success. According to Chapman, the real value in being a successful manager is that it gives you an incredible opportunity to have a positive impact on hundreds, or even thousands—perhaps millions—of lives. If you handle the job right, you can impact every employee, every customer who uses your products, and the employees of every supplier, as well as countless people in your communities, in a positive way. That—and not financial reward—is the thing of lasting value you can take away at the end of your career. Of course, the company has to make money along the way for you to fulfill this opportunity—and Chapman does it. He makes a lot of money for his company, which enables him to touch all of those lives in a positive manner.

To reach the excellent Barry-Wehmiller levels of performance, members of senior management first need the courage to get rid of the jerks. If 95% of the people in any organization show up every day, willing and often eager to do their job as best they can and to behave in a positive, supportive manner, then 5% of the folks have some other agenda. Just about every company has them—often perceived to be indispensable because of a tremendous amount of tribal knowledge in some aspect of the business or another.

Usually senior managers are blind to these folks because they seem to be very capable: walking encyclopedias of product and part numbers or the arcane details of some system or procedure. They thus come across to their bosses as quite valuable to the business. More often than not, however, they are the people who use that vast knowledge to argue why things cannot be changed, why other people's ideas and opinions are no good, and why they have to continue to be at the center of everything happening within the company.

One manager described these folks as being like the alligators and crocodiles in the river. They win the fight by dragging their prey down to the bottom where they are invincible. On dry land, they are not nearly so tough. The people in your organization who fit this mold are the ones who often won't engage in ideas—instead they want to pull the discussion down to detailed exceptions and the minutiae of the business where only they have the facts.

Those unfortunate enough to have to report to these folks cannot stand them because the only behavior they support is that of lackeys, and their peers tend to avoid them as often as possible. The great frustration in the organization is members of senior management's defense of the jerks, showering credibility and public praise on them.

The problem is that they are indispensable to senior managers but not to anyone else. There are usually lots of folks fully capable of getting the jobs done that these folks dominate—often capable of doing it better—but they are not allowed inside the tent.

Senior managers must recognize that they have a unique, skewed perspective. For one, people often behave very differently around the big boss than they do with their peers and subordinates. Also, what is an impressive command of detail to senior managers is usually common knowledge at lower levels. The difference is that most people do not feel compelled to impress senior management with that knowledge.

It falls on Human Resources to root these people out of the organization. Peer reviews, 360° performance appraisals, and "bottom-up" assessments are typical techniques. In one form or another, the people imbued with authority in the company have to be subject to formal feedback from their peers and subordinates, and senior managers have to put more weight on these feedback processes than on their own limited observations of the first-line management of the company.

No matter how it is done, the few in the organization whose agenda is a selfish preservation of power have to be rooted out. All of the platitudes about people being the most important asset are quite true, as are the ones about being only as strong as the weakest link. Detailed knowledge of how things are currently done is vastly overrated in most companies, and the willingness to learn and work hard tend to be equally underrated.

The most valuable person in the organization is one who can walk through the door and perform any job in the place—and is willing to do so. Of course, that person rarely exists, especially in a good-sized company, but the principle stands. The more jobs the people in the company can do, the greater flexibility the company has and the less is the need for specialized support people. Not only should people be able to perform a number of production jobs, but, more importantly, they should also be self-supporting.

It is a terrible waste of talent when, as is almost always the case, the best production people are moved out of the jobs in which they actually create value for customers—production—and are put into non-value-adding jobs inspecting parts, handling material, and supervising. Consider a young lady we know who has worked her way from an entry-level, shop-floor position to a fairly important role in cost accounting. When asked to take over the finance role in supporting one of the major value streams, she wanted no part of it because the job was "back on the other side of the wall"—within the plant instead of in the office area. This way of thinking is a very common culture and serves as a big obstacle to excellence. People have been conditioned to believe that the worst job in the company is the one that actually makes things—and the further from it they can get the better their job and the more important their role. Usually there is good reason for thinking that way: in most companies the further people are from production the more they are paid. This must change.

One necessary solution is to get rid of production support jobs altogether. Contrary to the gloom and doom about the education systems, workers in the United States, Australia, and Western Europe are far better schooled than their low-cost country counterparts, where the norm is a sixth-grade education at best. The superior education of your employees has to be put into play. It is hard enough to compete with the wages of backwater places without compounding your production folks' wages with unnecessary support. There is no reason people cannot get and keep track of their own parts, inspect their own work, and maintain their machines, and they certainly should not need a glorified babysitter to make sure they show up for work on time and stay at their tasks.

In a typical manufacturing company, the ratio of direct, production employees to indirect hourly labor is close to 3:1. In the best companies it is over 4:1 and often closer to 5:1. These companies accomplish significant elimination of non-value-adding labor cost by relentlessly training their people and setting expectations higher for people to be self-supporting.

The BAE System plant in Aberdeen, South Dakota, is perhaps the best example of just how effective the human resources function can be handled. There are two levels in the plant: the plant manager, and her 250 or so direct reports. The employees are almost entirely self-directed, working in teams that set their own flex-time hours and production assignments. The only way to get a raise is to learn a new skill, so there are few people who can only machine, weld, assemble, handle materials, or anything else. It typically takes a year or two to master one of the critical skills—mastery is certified by that person's peers—and a pay increase is granted. Employees need to move on and learn something else, getting started on the next year or two's efforts, to gain another pay increase. In the meantime, however, they are working in a production job that decreasingly needs anyone else to support it.

The plant, by the way, sets records for delivering missile launchers on time to the Navy and for continually reducing costs. It also has one of the thinnest human resources policy manuals around. The guiding principle is not rules but fairness. Fairness in most companies has been defined as treating everyone exactly the same—largely to avoid even a hint of discrimination. When most of the decisions typically made by managers in accordance with a rule book are made by an employee's peers, however, a whole new definition of fairness emerges. It has more to do with the true definition of the word *fair*. When a very good employee, whose spouse also works, has young children with the flu, in most companies they are on the horns of a dilemma, trying to determine which parent can better withstand the absenteeism policy violation. Members of management treat the slacker who needs a day off to recover from a hangover the same as an outstanding worker with sick kids often because they don't know the difference, and when they do it is to avoid being sued for mistreating the slacker. The peers know the difference, however, because they work next to them all day long, and that lawsuit is a lot harder to bring against a self-directed work team setting its own flexible work schedules.

The other ingredient is incentive compensation. Management concerns that employees left to their own devices—whether in handling and assuring the proper accounting for materials, inspecting their own work, certifying each other's skills, or having input to individual and group scheduling—will not live up to their expectations can be allayed by giving a significant portion of compensation in the form of bonus pay based on bottom-line results (e.g., delivery performance, value-stream profitability, quality levels). This aligns employee goals with management and the rest of the stakeholders. They are just as concerned with getting things right and making sure good employees are supported while bad ones are moved out because their paychecks largely depend on those results.

The bottom line in managing people is this: (1) get rid of the bad eggs and move the best people as close as possible to the line of fire where production and value adding are taking place; (2) train people relentlessly and pay generously for skills (and give that generous pay from what is saved by eliminating non-value-adding positions); (3) set high expectations for people to be self-supporting; and (4) treat everyone fairly instead of the same and enlist the help of every employee to assure that fairness is properly defined.

7

Channels and Chains

Everything that happens from the time someone digs the material out of the dirt, cuts down the tree, or otherwise pulls it out of its natural state until it gets to you is called the *supply chain*. Everything that happens after you until the final consumer buys the product and uses it up is called the *distribution channel*. These terms are just two ways of basically saying the same thing—the whole series of events required to go from birth to death of a manufactured product—but your suppliers think you are part of their distribution channel, whereas your customers see you as part of their supply chain. Whether you are a link in a chain or the controller of a section of channel, the important point is that you are not in control of much of anything. Rather, you are merely one step in a whole series of events and organizations, and your success is a function of how that whole thing performs. The sole judge of whether the channels and chains are successful is the end consumer, who sets the rules; you live or die based on how well you execute in conformance to those rules.

As fundamental as this might seem to be, or as much as managers want to write this off as some theoretical macroeconomic gobbledy-gook with little application to their real day-to-day work of making a profit, failure to understand and put into practice this basic idea is what causes manufacturers to lose customers and sales. It is not about you: what you want or need, what you think you do best, your "core competence," or anything else having to do with your business. Instead, it is about what the end customers wants, and you either deliver or don't.

Every purchasing person has long experience with suppliers or would-be suppliers attempting to dictate lead times, payment terms, quality procedures, and, of course, price. In fact, those aspects of the relationship between sequential players in the chains and channels are not up to either of them. The notion that the lead times, for instance, should be on your terms rather than the customer's is a very common example of missing the basic point of supply chain economics entirely.

Perhaps your customer is selling to Walmart, which drops orders via electronic data interchange (EDI) with the terms that the order must ship within 72 hours or be canceled. Or maybe your customer is an auto assembly plant with a 10-day rolling schedule. Maybe your customer is making missile launchers for the U.S. Navy with a set contract for delivery according to the schedule for the construction of a new Navy destroyer. When you think you can "negotiate" lead times greater than the terms your customer has to live up to, you are simply asking your customer to finance a disproportionate amount of inventory. For example, the customer ships to Walmart in three days, whereas you want two-week lead times. That means there are 11 days of uncertainty your customer has to cover with an investment in an inventory of your products. To the extent that it cannot make its own products in three days, it is already carrying an inventory of its own finished goods—and now you want it to tie up cash and floor space and eat the carrying costs of your product, too? No, this is hardly some vague macro theory meaning nothing in the day-to-day rough-and-tumble of buying and selling. It is right at the heart of why companies are often too willing to dump one supplier and move to another.

If your customer's customer pays in 60 days but you have a "policy" of selling on two 10-net-30 terms, you are doing nothing more than asking your customer to be your banker and to finance your investment in the supply chain.

The Navy is demanding a 5% annual cost reduction in its contract to build destroyers; retailers are in a battle to continually reduce prices to consumers; car companies are doing the same. Meanwhile, if you are sitting a mile up the channel or eight links up the chain, telling your customers that you *need* a price increase, then you are out of touch with this basic point and are setting yourself up for failure.

The old McCulloch Corporation,[16] well on its way to bankruptcy at the time and not too surprising to anyone familiar with it, once decided that Home Depot's policy of accepting customer returns for any reason was unreasonable. It unilaterally established a return authorization policy

---

[16] From author Bill Waddell's experience as VP of Supply Chain at the McCulloch Corporation, Tucson, Arizona from 1996 to 1999.

and had the gall to send a pad of preprinted forms describing the defect and justification for the returns to all of its customers—Home Depot included—informing them that a completed form had to accompany every returned product for McCulloch to issue a credit. A few weeks later, a skid of returned products arrived at the McCulloch warehouse in Tucson. Accompanying the skid was Home Depot's form telling—not asking—McCulloch that it was taking a credit for the product. Included was also one of its pads of preprinted return authorization forms. Across the back of the pad, on the cardboard backing, was scrawled in black marker, "This s**t don't work." McCulloch got the message—it did not really have the authority to decide what the return policies would be for that channel.

McCulloch personnel complained that some chain saws had been purchased, used once, and then returned even though nothing was wrong with them. Some guy had one small tree to cut down and basically borrowed a McCulloch chain saw from Home Depot to do the job. Was it unfair to McCulloch? Of course. But that is just the way it works in that supply chain. Home Depot—and, more importantly, Home Depot's customers—had determined that a "no questions asked" return policy was an essential element of the value proposition: the "reliability" part of the value troika of quality–utility–reliability. It took the risk away from customers who bought products they were often unfamiliar with or were not sure were right for the job. The appropriate response to this rule of the supply chain would have been to put the same terms to their suppliers—not to try to push back on Home Depot.

It works both ways, of course. Just like you cannot go to your customers demanding terms more lucrative than those defined by the end customer for the distribution channels in which you operate, you cannot tolerate suppliers coming to you trying to do the same. Suppliers that do are merely inviting you to go to their competitors and find one that will play by the rules of the chains and channels in which you operate.

By the same token, you can't get away with expecting more from your suppliers than the customers are demanding of you. When you get 10-day lead times and 30-day payment terms, if you think you can dictate one-week lead times and 60-day payment terms you are doing nothing more than urging your suppliers to change their focus to another distribution channel. This is the sort of nonsense the auto companies did under the guise of "Just In Time." GM simply pushed all of the inventory investment, risk, and cost back on its suppliers to make its own books look good, oblivious to the fact that the supply chain wasn't improved one iota. Doing this merely drove its suppliers to China and many to either bankruptcy or to strategies that lessened their dependence on the automotive industry. Toyota and Honda, on the other hand, have not forced shorter lead times on their suppliers so much as they have demanded that their suppliers develop the capability of producing in shorter lead times. The big difference is that Toyota and Honda were intent on driving inventory out of the supply chain all together, whereas GM didn't care about the supply chain so long as the inventory was not on its books.

Knowing the chains and channels in which you participate and picking the right partners is essential. There is no shortage of big companies like GM that either don't understand these basic points or don't care and freely throw their weight around, trashing suppliers and taking advantage of customers. When you accept purchase orders from a link in one of their supply chains—unaware of how fundamentally dysfunctional the overall chain is—you are setting yourself up for disaster. GM's bankruptcy and the demise of a bunch of big tier-one suppliers made the headlines, but they deserve little sympathy. They are the ones who chose to trash the supply chain for their own benefit, abusing everyone else down the chain and up the channel. Those who were forced to learn hard lessons were the thousands of small manufacturers—tier-three suppliers and the companies that sold to them—taking a beating or going under completely because they failed to look all the way down the distribution channel and see that they were hitching their stars to an "every man for himself" supply chain.

Customer–supplier partnerships have been a widely bantered buzzword in recent years and have generally been misunderstood and abused. This idea of knowing the rules of the supply chain as set by the end customer and becoming very, very good at executing in conformance with them is what partnerships are really all about. A chain or a channel of true partners is one in which all of the participants understand and share in the investment, risk, and effort to meet the end customer's definition of value. The more typical chains and channels are the ones populated by a bunch of mercenaries continually looking to get a few more pennies from both customers and suppliers to make themselves more profitable in the short-term, with little regard for the impact on anyone else and even less regard for the impact at the end of the channel where real customers reside.

One pretty good dictionary definition of the word *partnership*[17] is "a relationship between individuals or groups that is characterized by mutual cooperation and responsibility, as for the achievement of a specified goal." That gets at what a supplier partnership is really all about. Expecting a supplier to enter into a "partnership" and mutually cooperate and accept responsibility for the specified goal of making your company more profitable regardless of the impact on its supply chain or the supply chain in total is not any kind of a partnership. It will not work or withstand long-term stress. Only when the specified goal is maximizing value for your customers can a partnership with your suppliers have validity.

A big difference between the excellent and not-so-excellent companies is that the good ones are selective about their customers, whereas the bad ones jump at every opportunity to sell anything to anybody. One contract manufacturer we know has a 10-point assessment that all members of senior management contribute to in evaluating a new customer. The list encompasses things like the culture and strategy of the potential customer to make sure there is a good fit between the two companies. This manufacturer is trying to get at this definition of partnership: will this new customer be a fair, honest partner in the effort to continually improve? It is not looking for easy customers; in fact, one criterion is whether the new customer will put pressure on the company to get better in areas in which it knows it should get better. The excellent companies can cite examples of declining sales opportunities for reasons other than the creditworthiness of the potential customer.

It is not easy being a supplier to SC Johnson, Wahl Clipper, Parker Hannifin, or Boeing, but it is not easy being them either. They take care of some pretty demanding customers and operate in some high-stress supply chains and distribution channels. They expect no less from their suppliers than they expect from themselves, but they expect no more either. A manufacturer has to know the chains and channels in which it chooses to participate, and it has to know it all the way to the final customer. You have to be aware of the rules and make a conscious decision to play by those rules—or to find another channel in which the rules are more to your liking. Attempting to do otherwise is asking the good potential partners— both customers and suppliers—to sooner or later kick you out of their sandbox and find someone a lot more farsighted to play with.

---

[17] http://www.yourdictionary.com/partnership

8

You Gotta Go with the Flow

While it would be helpful if the sales and marketing community and the operations folks could get on the same page and decide whether we should call this series of activities a chain or a channel, we have to give them credit for both coming up with metaphors that denote continuity. They are all talking about the flow of materials all the way from the raw state to the final consumption of it in the form of some product someone will use to destruction. It seems reasonable to assume that how well that flow proceeds through your factory from the receiving dock to the shipping dock would be a big driver of how well you do.

In fact, all of the really big leaps in manufacturing have been big improvements in flow. Henry Ford's assembly line was nothing if not the epitome of how a radical improvement in flow drives a radical improvement in profits. The assembly line where the finished cars are put together is a big deal in the eyes of the public and the press, which know next to nothing about manufacturing. The professionals, however, could see that putting an incredibly huge and complex machining operation, that makes engines and transmissions, in continuous motion, was really the heart of it. He didn't put cars together in a matter of a few days; he took raw iron ore to steel to machined components to engines to cars in a few days—and made a staggering amount of money by figuring out how to do it. When Ford uttered the often scoffed-at line, "The customer can have any color he wants, so long as it is black," he was not only demonstrating an appalling lack of concern for how his customers define value (which would ultimately do him in), he was also talking about flow. The cost of a Model T was ridiculously low because of the ridiculously high rate of flow he achieved. Stopping the flow to change over to another color would have killed the giant cost advantage he had achieved. High flow means low cost.

Before developing into the ideas of Lean manufacturing and "*Kaizen* Kowboys" who want to come into your factory and lead a project laden with mysterious Japanese terms about *senseis* and *kaikaku*, the folks at Toyota described their extraordinary manufacturing approach as the continuous compression of time and space—the time it took to flow through their radically smaller factories.

Before Six Sigma[18] was hijacked by consultants wearing odd-colored belts, it was the driving force behind Motorola's incredible manufacturing success in the 1980s. Its abandonment of those principles was the energy behind its utter failures in manufacturing in later decades. At the outset Motorola was driven by the mantra, "The best quality producer is the shortest cycle time producer, and the shortest cycle time producer is the best cost producer." It was all about eliminating the variations in flow—the things that caused products to stop moving. Sitting idle, the fixed costs were rolling on without products moving in and out of the factory to absorb them. The impetus behind Six Sigma was exactly the same principle behind Ford's assembly lines and Toyota's Just-In-Time theory.

The idea that a factory should be broken down into departments—machining in one corner, molding in another, and assembly somewhere else—in the middle of the flow in from the supply chain and then out through the distribution channel defies logic. The justification usually has something to do with direct labor: either getting all of the similarly skilled people together so they always have something to work on, or putting them all under an overseer who knows all there is to know about his or her trade to prevent them from buffaloing a boss into unjustified slacking off. Regardless, any justification—especially one aimed at optimizing the 5% or 6% of cost direct labor typically represents—pales in comparison with the penalty for bringing supply chain and distribution channel flow to a halt.

Management guru Peter Drucker wrote in his omniscient article "The Emerging Theory of Manufacturing"[19] in 1990 that the factory should be "little more than a wide place in the manufacturing stream." Warehouses and inventories may be "necessary imperfections" in the stream from the

---

[18] From author Bill Waddell's experience at the Motorola Institute in 1990 and 1991 taking courses on Six Sigma, quality control and supply chain management; as well as Bill Waddell's conversations with former Motorola Purchasing Director at the time of Six Sigma's development, Kenneth J. Stork. Six Sigma is a registered trademark of Motorola, Inc., Schaumberg, Illinois.

[19] Drucker, P.F.; *The Emerging Theory of Manufacturing*; Harvard Business Review; Number 90303, May-June 1990.

beginning to the end, but nonetheless they are clearly imperfections. They are hardly something to be designed into the factory to figure out how to squeeze a few more pennies out of direct labor.

The theory of constraints, offered up by Eli Goldratt in *The Goal*[20] more than 20 years ago, is an enormously powerful manufacturing management tool, not just for factory efficiency but also for the whole business. Goldratt and his various business activities have done as much harm as good in getting that point across. Some folks call themselves "Certified Jonahs," shrouding the straightforward idea in mystery, much like the Lean *senseis* and Six Sigma black belts who have a vested interest in making the principles much more complicated than they really are to make the case that you need to hire them and pay ridiculous fees to avail yourself of their knowledge. Nonetheless, throwing the baby out with the bathwater because you rightfully don't buy into the expensive and over-the-top proposals these consultants offer would be a serious mistake. Just because you don't need a Jonah from the Goldratt Institute doesn't mean you should not learn and keep a keen focus on the theory of constraints.

In a nutshell, the theory is that there is a bottleneck, or constraint, in every process and that the constraint serves as the gateway for all of the flow through the plant. Foul things up and lose production at the bottleneck, and you have lost the production forever, plus the fixed costs that continued to roll on as the bottleneck remained idle. Improve things at the bottleneck, on the other hand, and you have improved things for the entire factory, not just for the labor cost at the bottleneck. As Goldratt says, "An hour saved at the constraint is an hour saved for the system; an hour saved anywhere else is a mirage." If you have an operation that can make a part a minute, everything before that operation and after it is wasting time if it tries to produce any faster than one per minute. You can make a machine or a person look very productive for short periods of time producing faster than the bottleneck, but, at the end of the day, nothing is going to leave the factory any faster than the constraint. Those apparent efficiencies are Goldratt's mirages.

The big results from efficiency improvements and capital investments at these nonbottleneck operations are Shakespeare's "sound and fury signifying nothing." Most of the time, when manufacturers express

---

[20] Goldratt, E.M.; *The Goal: A Process of Ongoing Improvement*; Gower; Surrey, UK, 1993; for additional information on the Theory of Constraints and Throughput Accounting see the Goldratt Institute Web site http://www.goldratt.com/

frustration at having sent Lean experts around running *Kaizen* events at great expense to the consulting budget or have brought in Six Sigma black belts to make big changes, they have done so without really understanding Goldratt's point. The tools of Lean manufacturing and Six Sigma were all designed to improve flow, not to reduce isolated costs or to hammer on direct labor expenses. Using them on something other than flow or on flow without considering the bottleneck is pointless. Improving some isolated section of factory flow brings virtually nothing to the bottom line if the improved flow is simply going to pile up higher and faster when it hits the constraint.

So the overarching objective of the factory is to continuously accelerate the rate at which stuff flows in the back door and out the front door, and knowing what limits, or constrains, flow is obviously a pretty important bit of information to have. Keep in mind, however, that manufacturing operations and machines might not be the only constraints. In more than one factory we have seen, the constraint is in nonproduction areas. The machining business constrained by the engineer who has to review customer drawings and turn them into shop drawings and machine programs is fairly common. Order entry and customer credit approvals often limit manufacturing flow.

You should keep in mind that there will always be a bottleneck. Take that old operation that could only produce at one per minute and automate it down to one every 10 seconds; although you have made a big improvement to an isolated cost, you have merely moved the bottleneck somewhere new. Now some other machine constrains the factory to one every 45 seconds and must be identified and optimized just as diligently as the old constraint. This is important to realize before you invest in the new whiz-bang automation. In this example, the return on the investment in automation will not be the 83% reduction from one per minute to one every 10 seconds it appears to be. It will be a 25% improvement from the old one per minute to one every 45 seconds—the rate at which the new constraint will support.

This all seems quite simple and quite obvious: the whole family gets out the door and to grandma's house for Sunday dinner as fast as the slowest family member gets ready and out the door. Like every point we have made when it comes to manufacturing excellence, the key is to clear away all of the clutter that makes simple points like this so difficult to see. Get rid of most of the complicated mathematics having to do with discounted cash flows and internal rates of return, just go out and find the bottlenecks, and your investments in capital improvements will pay off handsomely. Send the Lean teams and Six Sigma experts after your constraints, and their efforts will look good in actual practice, not just on paper.

There is a very real economic significance to changing management's thinking to a flow orientation. Early on we talked about companies like SC Johnson and Wahl Clipper, which have long track records of stable employment. Toyota's lifetime employment philosophy is legendary. Even with all of its current, self-induced problems, with the exception of the folks who worked as part of its minority ownership of the New United Motor Manufacturing Inc. (NUMMI) joint venture with General Motors and a relative handful of temps at its Texas truck plant, Toyota still has not laid off any employees in the United States. Many manufacturers—probably most manufacturers—say they would love to do the same, but they cannot see how to avoid having to lay people off when their sales volumes go down. The answer is pretty straightforward: make sure sales volumes don't go down.

In Chapter 17, focused on strategic pricing, we will detail how to accomplish this, but for now keep in mind a couple of basic principles. For one, the best way to stabilize the workforce and sales volumes is to control the upside. When sales team members are let loose to sell anything to anybody any time and all of their metrics are based on the assumption that more is always better, you end up with a factory and a company teetering on chaos. The behemoth of a factory and its long supply chain is expected to chase sales volumes up the peaks and down into the valleys, continually overusing and then underusing capacity, lurching from laying off to rampant hiring, from all-out expediting to inventory reduction.

Running factories at their flat-out top speed and then slamming on the brakes is no more economically efficient than is the fuel efficiency resulting from driving your car in the same manner. Just like your car operates at its best from the steady pace of highway driving, your factory thrives on steady and stable levels of demand. Toyota's old philosophy—one that most "Lean experts" never learned or quickly forgot about because it doesn't fit the traditional model—was, "Factory first." It didn't mean the factory was more important than anything else; rather, it meant that the objective is not higher sales but maximum profits. Running the factory at the optimum level means optimum leverage of fixed costs. It means there is a sweet spot, usually in the 85–90% capacity use level, at which the factory is at its best overall cost. The most profitable the company can be is when it operates for sustained periods of time at that level. Continually increasing factory capacity at a steady rate and sales at the same rate means sustained, very profitable growth. Simply ramming more volume through the factory is not particularly profitable.

When a company makes a commitment to stable employment similar to that of SC Johnson, Wahl, and Toyota, it converts payroll to a fixed cost. At that point, everything except for materials is essentially fixed, so the mathematics become simple: get more volume across that fixed-cost base. That is all Ford did. It was no complicated theory. The factory cost about $200,000 a day to run plus the materials. Spreading that cost over 1,000 cars a day reduced the cost per car from a 950-cars-a-day rate by about $10 each. There was nothing more complicated than that behind the thinking. The same was true at Toyota, and it is how the other stable employment companies think. It is not a vast new theory of manufacturing accounting; instead, it is a rejection of the theories it takes cost accountants years of education to derive.

Becoming a flow-driven manufacturer, rather than the sort of manufacturer with disjointed operations scattered around the factory all focusing on isolated ideas of increased labor and machine efficiency, is a huge step forward. It is also a necessary step forward if you want to adopt the basic cultural principles of making a commitment to your people compatible with high profits. They go hand in hand. Stop all of the laying off and hiring, stabilize the factory employment, then focus on how you can get a steady and steadily increasing flow of sales and production across the plant. It's not easy to do, but it is not complicated. The rest of this book is directed at spelling out how you can go about making that happen.

Section II

Managing Differently

We have talked quite a bit about what is wrong with the way things are done and have expressed some radically different ideas, such as how you approach people and focus on value. At the end of the day, we are talking about a completely different approach to managing. It should come as no big surprise that management has to undergo some radical changes if the business is going to operate at radically higher levels.

It is actually some rather outrageous wishful thinking to believe what most of the consultants tell you about Lean manufacturing, Six Sigma, ERP, customer relationship management (CRM) software, and other canned fixes transforming all that ails manufacturing. They will have you believe that all members of management can pretty much operate as they have all along and that a little software or a project or two to rearrange the furniture in the factory will fix everything. Those might be attractive ideas, but the harsh reality is that the results you are getting are the product of how you manage the business, and if you want radically different results it will come from radically changing management.

Look at things from the point of view of someone supervising the factory floor. Human resources (HR) hires the people working on the floor according to some hiring criteria established by management and compensated according to management policy. He gets the machines justified however management decides new machines will be justified. He is given materials purchased from suppliers management chose, and the materials arrive (or don't) based on the scheduling systems and inventory schemes management put in place. Management hands him the schedule, the product specifications, and the delivery requirements. Management actually puts the factory in a very small box without much latitude over anything: the idea that management can basically say, "We aren't going to change any of these things—we are going to keep doing everything the way we have always done them, providing the 3 M's of men, material, and machines the same as always, but you should do cartwheels and handstands inside your little box with this new software and these Toyotaesque techniques and take the company to a whole new level of performance all by yourself." Then, a year later, management decides that whichever fad they pursued must not be effective, and they go off in pursuit of the next fad, never stopping to think that the real problem is in how they are managing.

You get what you manage. If you want the business to get different—better—results, you have to manage differently: better. If you want manufacturing to get different output, you have to change the box you have put manufacturing into.

A big part of the problem with ideas like Lean manufacturing and the Toyota Production System is that they were introduced to the United States and the rest of the world by a bunch of nonmanufacturing people. The academics and writers who first went to Japan to see it in action weren't really manufacturing people, so they didn't ask the right questions. They knew only what they could see—which was a lot of different factory stuff: *Kanbans* and 5S; standard work and *Kaizen* events. A grizzled old manufacturing person would have asked a lot of questions of the folks at Toyota: mostly, "How do you get your accountants to go along with all of this stuff?" That would have opened the door to a world of things about how Toyota and others are managed.

A similar failing took place with guys like Jack Welch[21] and the others who jumped out in front of the Six Sigma parade. They never asked the questions that immediately come to mind largely because the idea that management—them—could be the real problem would never cross their minds in a million years.

The simple fact is that the companies doing the impossible very successfully are doing so by managing their businesses very differently. Most of the excellent companies are privately held and, as a result, not the least bit concerned with what Wall Street or anyone else thinks of their performance in the short-term. Because privately held companies are much more cash driven—like the entrepreneurial start-ups—they already have a very different view of accounting.

It also comes easier to companies like SC Johnson. It is no surprise that Ford is the most successful American car company. Like SC Johnson, it has (1) its family name on the product and (2) a lot of family input in how the business is run, even though both it and SC Johnson are publicly traded companies. The ideas of value and commitment to employees, suppliers, customers, and communities come a lot easier when it is the family name on the pink slips and in the newspapers when you decide to close plants, outsource, and devastate communities. So the decisions that emerge from these companies are much different from those at most of the big public companies.

The next several chapters describe the differences in the nuts and bolts of management—how the excellent companies actually do the day-to-day work of running the place that results in better outcomes. The biggest difference centers around accounting, however. That is the lynchpin. If you insist on measuring yourself the way Wall Street and the Internal Revenue Service (IRS) measure you and refuse to believe that standard costs and numbers derived from generally accepted accounting principles (GAAP) are wrong, you will never get there.

The entire integrated management infrastructure you have in place now—whatever it is—is driven to optimize financial performance, however you choose to define financial performance. If you think inventory is really an asset, you have systems and procedures in place to "optimize" inventory rather than eliminate it. If you think that direct labor is a critical variable cost that actually drives the overheads allocated on the basis of it, then you most likely spend a lot of time worrying about head count and have metrics myopia about reducing it and squeezing more out of it. If you don't think cash flow is as important as book profits, then you don't have management processes focused on product flow since cash flow is directly related to manufacturing flow.

The companies that do their accounting differently—the ones that pay short shrift to book profits and care about real money (i.e., cash)—do everything differently. They care quite a bit about production flow and see inventory as a big negative. As a result, they approach their supply chain and the purchasing function in an entirely unique manner. They don't worry about the 5% or 6% of their costs direct labor represents nearly so much as they worry about costs that are not directly resulting in increased selling prices. They know that direct labor is generally creating value and is more a good thing than a bad thing—at least so long as it is making things that get shipped. Making things to sit in inventory is bad all the way around, as these folks see it.

That radically different management infrastructure is the subject of the rest of this book. We will touch on each of the critical areas of management and explain the approach taken by the excellent companies. You really have to wrap your mind around the accounting issue for it to make sense, however. If you are stuck in GAAP, none of it will click. Likewise, if you change only one thing in how you run the business, change your accounting system as far as it affects management. If you do just that one thing, all the rest will fall into place on its own.

The theme has been simplicity, and that is exactly what this stuff is. If you change to a simple, straightforward accounting process, one by one all of the wasteful, complicated management systems you currently embrace will sooner or later fall by the wayside as the power of focusing on the basics becomes more and more evident.

9

Getting Sales and Marketing into the Game

In most companies that claim to have launched grand improvement strategies (i.e., some version of Lean or Six Sigma), the sales and marketing function is nowhere to be found. They are pretty much still off doing what they have always done without much regard for what they, and the rest of senior management, typically view as strictly operational concerns. Unfortunately for the companies that roar down that path, the words of English poet John Donne apply to businesses as well as people: "No man is an island, entire of itself." You can't launch any operational strategy and have the sales and marketing team members off doing their own thing and expect the business as a whole to gain much by it. At best, you get organizational conflict and confusion; at worst, the whole place comes unglued.

With all we have discussed concerning the essence of value to customers and how you cannot possibly succeed unless you know very clearly exactly what value is and about the supply chain and distribution channels being the definers of what you must do to succeed, it should be clear that sales and marketing have a far greater role to play than merely finding more people to buy more stuff.

At the outset of this book, we briefly described value streams as cross-functional groups whose mission is to take customer orders from start to finish and whose intent is to deliver the maximum value for the customers, however they define it. Without sales and marketing tied tightly to both the customers and operations—in effect serving as the critical link between the customers and the value-creation activities—there is virtually no chance of success.

In most companies, when all of the eloquence natural to sales and marketing folks is stripped away, the objective of these teams is to sell anything in the product line to anyone who will buy it in the greatest quantities they can write in the sales order. As far a target goes, that is an impossible one for the company to align itself toward and hit. It is true that failing to plan is planning to fail. Planning is a necessity for success, but that doesn't mean it needs to be complicated. There is little call to run out and hire a consultant.

Developing a sales and marketing strategy is difficult only if you don't have the necessary, basic information. That includes knowing what distribution channels you operate in, who the customers are, and what they want and need. You also need to know who your competitors are and how they go about meeting those needs. The plan is nothing more than a basic scheme to lay out how you want to meet those customers' requirements better than the other guys do. It seems simple enough and it is—*if* you have that information.

For each channel, you lay out what you are going to sell and to whom, what the rules are for success in the channel, and how you are going to meet or exceed the rules better than the other guys.

Even if the information is unavailable, you still have to plan, using educated guesses to fill in the blanks and then embarking on the never-ending process of continually learning more about the customers, competitors, and the value proposition needed for success. The first order of business is to identify the customers; in just about every business, one size does not fit all, so you have to understand the different markets and channels you serve.

As we laid out in the previous chapters, the center of things is value: creating it as best you can and eliminating the waste that doesn't add to it. Different customers will define value differently even though they might be buying very similar products from you—maybe even exactly the same product from you.

You might be selling to consumers and to professionals in much the same manner as the Wahl Clipper. The consumers have a different set of requirements for your products than do the professionals who want to use your product to help them make a living. At the very least, they have different views of the utility factor. A consumer will use it occasionally, whereas a pro might use the product every day.

Like Parker Hannifin, you might be selling to the automotive sector as well as to the military. The automotive industry is very cost driven and wants no frills and the lowest possible cost, whereas the military is far more concerned with reliability: how long the product will hold up under tough use. These are matters of degree. Both markets are concerned with cost and reliability, but the shading of their relative importance is huge.

Even when you are selling the same product to very similar customers, when some of those customers are domestic and some are overseas there is a sharp difference in how they define value. The domestic customer is apt to want frequent, small shipments with short lead times and all of the interface handled electronically by EDI or other means, whereas the export market allows for longer lead times but requires container loads, mountains of export documentation to facilitate foreign customs, and unique financing involving letters of credit.

The point is that you have to break your customers down into the various, logical groupings based on how they define value. Generally this means that you have to break them down by channel since the end customers in any particular channel tend to define value alike. This is not always true, however. Sometimes there is one big player that sets its own unique rules. Walmart, for instance, with more than 40% of the retail sales in North America, is quite apt to be a channel all by itself, with the rest of the big-box retailers forming another distribution channel. In subsequent chapters we will discuss how to align the factory and the supply chain to deliver that value, but it all begins with getting a clear idea of the various definitions of value you have to fulfill.

Within each channel, the next step is to identify exactly who the specific customers are and what products they need and to define their value needs in every aspect. What price do you need to hit? What lead times? What documentation? What payment terms? What quality specifications and documentation? What warranty and after-market support?

The objective is always to grow market share, and quite often this means taking on a competitor head to head. The strategy you come up with can be a very predatory one, aimed directly at a single competitor with the intent of undercutting that company at every turn and driving it out of the market. There is nothing wrong with this plan so long as you know what you are doing and are prepared for however that competitor will respond.

At the other end of the spectrum, the strategy may be one of solving problems the customer doesn't even know it has yet. Most companies that offer to put inventory in consignment are not doing so for altruistic reasons. They do it because it gives them the inside track to your business and the chance to nose around and be the easiest solution to whatever problems arise for your business. One such company used this technique to grow its percentage of its largest customers' purchasing spending from about 3% to over 10%. Quality Screw & Nut Company (QSN)[22] from Chicago (acquired by and now part of Anixter) realized that eliminating the hassle of purchasing the nickel-and-dime parts it sold was as valuable to its customers as the screws and nuts it provided. Its business model provided a consignment inventory along with an on-site representative, whose very part-time role was to manage the inventory; the real objective was to take on as much of the "C" item sourcing and purchasing as the customer wanted to offload to QSN. Had it not understood the broader definition of value, it would have been just another outfit peddling hardware and butting heads with cheaper China sources—and most likely losing the head-butting match.

Other companies do the same sorts of things (i.e., freely provide services above and beyond the basic requirements of the purchase order for not too well-disguised ulterior motives) most often to be sure their folks are hanging around and freely available when new products are being designed. They realize that their expert knowledge of the components and materials they provide can be of substantial value to the engineers trying to come up with new and improved products.

One structural steel company we know is privately owned and has lots of cash but is run by financial people with little sense of the company's customers or how it can provide value. When the global credit crunch hit, the people in charge put in ever-tighter credit controls in a well-intentioned effort to protect the company from having some of its smaller customers go belly-up. Its chief competitor did the opposite. Knowing that these small construction guys were going to be especially hurt by tightening bank credit, the competitor offered all sorts of financing schemes to make it easier for the little guys to buy steel when the banks were unwilling to help. Of course, it tried to be smart about who it gave credit to, and of course it got burned a bit when it bet on the wrong horse, but it grew its share of the market so radically that it was a small price to pay. And it will

---

[22] From author Bill Waddell's experience with Quality Screw & Nut while working as VP of Supply Chain at McCulloch Corporation from 1996 to 1999.

reap the benefits for a long, long time to come: all of the contractors came to see it as a "partner" while they viewed the tight credit company as a "supplier."

It is all a matter of understanding the market and the customers and knowing what they really need in the broadest sense. Once you do, it is a straightforward matter of laying it out for each channel and then aligning the resources of the company behind the plan.

10

Value-Stream Structures

Having talked about flow, the need to link the inbound supply chain with the outbound distribution channels, and the critical nature of identifying how the customer defines value—and then going about meeting that definition—it is time to organize the company around making all of these lofty ideas reality. That means restructuring the whole place into value streams: knocking down all of the functional silo walls, with no more departments based on what kind of work people do. Everyone has to do the same kind of work—create value for customers. Functional departments never were a particularly good idea, and they most certainly are not an effective way of getting much of anything done that a customer sees as helpful.

A value stream is the series of activities that link each distribution channel you serve with its inbound supply chain. A value-stream organizational structure is one in which all of the folks who are needed to support the flow along the value stream work for one boss: the value-stream manager. There is no engineering department or supply chain department, no purchasing department or quality department. All of the people who used to work in those groups of like-skilled people are redeployed to the value streams, and no matter what their technical skills they are assigned responsibility for creating the maximum value for the customers of their particular value stream. They are to work cooperatively to make sure that the value stream has all necessary skills at hand to continually make things better.

To illustrate the point we cite the following anonymous but true story:

> The engineering leader of a value stream in a consumer products manufacturer left a meeting in which the key metrics for the value stream were reviewed with senior management. He was pleased but still a little surprised at the changes in his job, his outlook, and the level of involvement he had,

less than a year after the transition from functional departments to value streams. His boss was no longer the director of engineering but now was a value-stream manager who had come out of the supply chain management group, apparently with only limited technical know-how. Where his previous peers had been three or four engineers with qualifications similar to his own, he now worked alongside a manufacturing leader, a marketing leader, a supply chain leader, and a value-stream analyst—people whose job content was as mysterious to him as were the nuances of his engineering expertise to them.

He couldn't help but be pleased. Senior management had just showered him with praise. The value-stream management team was a big step closer to reaching its performance bonuses because the value stream had made enormous strides in reducing inventory, one of five key performance indicators (KPIs) upon which value-stream success was measured. And the one KPI that had shown the least improvement was inventory reduction until the team had asked him in to take the lead. It turned out that the biggest opportunity to reduce inventory, along with corresponding costs and floor space, was to reduce component proliferation. Over the years, dozens of variations of just about each component had been created, many with little difference between them.

In the past, supply chain people had complained from time to time and occasionally had asked engineering department members to look at whether it was necessary to have so many different part numbers; however, those requests were seldom reflected in the engineering department goals and objectives, and inventory reduction objectives were nowhere on the engineering radar screen. Supply chain was measured on inventory turns. Engineering's goal was to get products designed on time and at the right cost target. If the goal was to get the product designed at a tough-to-reach cost of $5 per unit, engineering was not about to work in an existing component that cost 10¢ when a new part could be designed and procured at 8¢. And there was no time in the design schedule to go back to all of the products calling for the 10¢ part and redesign them to take the new, cheaper component. That was a cost-reduction project someone else would work on someday when and if there were any pressures to cost reduce those products. That is how it went on for a very long time.

Everything changed once the value-stream structure came into place. There were no "engineering goals" or "supply-chain goals," just business goals based on the KPIs that senior management had laid down. The value stream had all of the cross-functional talent and authority to do just about whatever it took to reach those KPI goals. Inventory was one of those goals the value-stream team had decided to focus attention on, and the engineering leader found himself cleaning up hundreds of part numbers with enormous impact. Profit was increasing, product costs were dropping, and inventory was dropping even faster.

Better yet, from the engineering leader's perspective, he was getting support he had never imagined in his product development efforts. His value-stream peers were driving supplier support, making machine time available for prototyping, and giving him technicians and quality inspectors whenever he needed them. His projects were coming in on time for the first time in his career as he no longer had to fight and negotiate for resources.

Breaking up all of the old functional silos and putting people into cross-functional value streams, in one fell swoop, changes everything for the better. It aligns resources on the things that really matter, gets rid of all of the departmental squabbling and politicking, and focuses the entire organization on creating value for customers. It makes excellent business performance possible and, just as important, sustainable, whereas simply value-stream mapping as part of some Lean *Kaizen* event or Six Sigma project usually do not.

Creating a value-stream map and defining a wonderful future state and then sending all involved in the process back into their current functional (or more accurately, dysfunctional) departments rarely result in significant change for obvious reasons. The suboptimal current state did not just happen; it exists because people perform to their job descriptions and the priorities given them by their supervisor. They work toward meeting departmental metrics and the accomplishment of strategic objectives for their functional group. The value-stream engineering leader in the previous example had been in a number of meetings in which part-number reduction and concerns about inventory levels had been raised. He and his engineering peers had agreed to look at it with good intentions. The meetings ended, however, and the higher priority engineering projects often ran behind schedule; in the end there just wasn't time to work on those "extra" projects, so they kept dropping down on the priority list until they dropped off all together.

The problems with converting to value streams are purely cultural and psychological. Everyone gets knocked out of their comfort zone. Managers who thrive on "command and control" go into a panic and can foresee only gloom and doom. No one doubts that cross-functional leadership is much more effective. It is a bit ironic that the traditional functional organizational structures were originally based on military command models. Today the military has moved radically to joint services structures because it is clearly much more effective to have such cross-functional integration, while many businesses still cling to their outdated ideas of the need to command and control everyone.

In a value-stream structure the old functional leadership stays in place; it just no longer has line control over the bulk of the employees. Workers are transitioned to the leadership of value-stream managers. The role of senior leadership switches from operational control and direction to providing broad, strategic direction. People in leadership are detached from the day-to-day running of the business to a role that both enables and requires them to look further down the road. They become technical consultants, teachers, and mentors, looking over the shoulders of their old functional employees and the new value-stream managers and providing advice and guidance as they find their way. Finally, they set policies, to the extent that policies must be set. Typically the value-stream teams are kept on a fairly short leash until they have settled in and demonstrated that they have things under control.

For the senior management team to relinquish control and turn things over to the value streams requires a high degree of faith in people. More importantly, it demands that all senior managers have faith in each other. After years of irrational, functional bickering it can easily be cause for concern. Salespeople may well have complained long and loud about the need for more inventory. This may have been an easy complaint to make, especially when they have had no accountability for inventory levels and finding space; counting it and paying for it have not been their problems. The supply chain manager may well lose sleep worrying about what will come of inventory when value streams laden with sales and marketing people have authority over inventory levels. On the other side of the table, production and supply chain teams may have been complaining for years about poor forecasting and unreasonable lead times, urging policies that customers be told no when they order something not in the plan or with insufficient notice. A degree of nervousness on the part of the sales manager would be justified when the individuals who have suggested such ideas—again, without accountability for the results—are put in control. Members of senior management need to have confidence in the people they have hired and developed over the years. Employees will do the right things if they are given the proper direction from the leadership team, and, just in case they start to stray, we will assure sufficient control over the value streams to enable them to quickly get back on the right track.

The critical issues in creating value streams are determining its scope and alignment and staffing. Just about every company initially misses the point in both areas, as it tries to apply old functional, command-and-control, cost-reduction principles to structuring its value streams. The driving principle, as we have discussed, is optimal value to the customers.

Most companies make more than one product type, although there are some commonalities among them. Wahl makes clippers and trimmers; Andersen makes windows and doors. Parker Hannifin makes dozens of related products that are similar. Most companies sell into more than one channel, perhaps into more than one market. Andersen, for example, sells both doors and windows to dealers as well as to big-box retailers such as Home Depot and Lowe's. Parker Hannifin sells many of its products to both commercial and defense customers. The first reaction by most companies is to view value streams as some sort of glorified cost centers: put all of the windows in one value stream and all of the doors in another to get the greatest manufacturing efficiencies and lowest costs. This misses the point.

The objective of restructuring into value streams is to align the company with each significant supply chain in which it participates—that is, by channel or customer type. Create one value stream to provide the maximum value to dealers and another to focus on big-box retailers; one for commercial customers and one for defense contractors; one for the domestic market and one for exports. Each supply chain has its own critical considerations, and the objective of the value streams is to make sure that the manufacturer is aimed directly at satisfying the customer's needs.

Each channel, and sometimes each major customer, has its own requirements regarding normal quantities and lead times, shipment documentation and labeling, order patterns and methods, required documentation, packaging, quality controls, and dozens of other variables. Every one of these has an effect on the flow of products through the plant and the flow of work through the production support areas. In the current credit crunch, each channel may have different requirements and constraints in financing purchases from you. The goal is to align the value streams to assure that all of those detailed elements of meeting the customers' needs are executed exactly and at the lowest total cost.

Best Buy or Target may submit orders via EDI with very short lead times for large quantities of a narrow range of items. The dealers may submit small orders via fax machine with longer lead times for small quantities of a wide range of items. The processes in place to optimize one set of customer requirements are quite a bit different from the processes needed to take care of the other. The gains in terms of total cost and long-term value resulting from excellent customer satisfaction always outweigh any manufacturing inefficiencies resulting from having to make two similar, perhaps even identical, products in two different value-stream locations. It is the weakness of standard costing that hides the overall cost–benefit stemming from value streams, which is a problem we will overcome in the next chapter.

The scope of the value stream encompasses customer interaction from start to finish. Every function that possibly *can* be taken to the value-stream level *should* be taken there. The only exempt functions are the few that truly must be performed but cannot be attributed to any one customer channel or another—financial accounting for external reporting purposes and supporting centralized computing services, for instance, and the purely "blue sky" research and development (R&D) functions.

The operational, engineering, and customer-related functions are typically easy to divide into value streams, but a number of others may not be so easily broken down. There may be two people handling product configuration tasks—engineering change and product change notices—and four value streams. Some training is going to be needed to convert two full-time jobs into four part-time jobs. The approach must be to keep the jobs out of the value streams and in their functional areas until they can be adequately defined and the people trained. Conversion to value streams is quite often a long transition that spans weeks, months, and sometimes a year or more. The important thing is to recognize that the transition has to happen and to have efforts in place to eventually have the value streams completely self-sufficient.

Along similar lines, it is not unusual to find that the factory has a shared resource or two (e.g., giant equipment "monuments" like paint lines or heat treat furnaces) that cannot be split and would be prohibitively expensive to duplicate. There is no avoiding having to live with a shared resource. One value stream or another has to "own" the monument and provide the use of it to the other value streams. As with splitting the jobs, it is more important to recognize the need to ultimately resolve the conflict and, when the equipment is eventually replaced, to do so with multiple and appropriately smaller versions that can be controlled by each value stream.

The next big challenge is to staff the value streams. Here again, companies often fall into the trap of thinking in a traditional manner, believing that the only person who can supervise the engineer in the value stream is a more senior engineer and that only a wiser and older salesperson can provide direction to a newer sales rep. The worry is that there will be no one to fill the value-stream manager positions if they have to be more qualified in every area than the functional leaders in the value stream who will report to them. Value-stream managers are not mini chief executive officers (CEOs), however. The leaders of each functional area within the value stream are assumed to be fully capable of looking out for their area of expertise without the need for someone to direct their every move. Technical support or guidance will be provided by the senior manager who used to manage that function. The task of the value-stream manager is to be the builder and leader of a team of people with a wide range skills and knowledge but with little familiarity with the knowledge of the others on the team. Leadership and communications skills are the most important attributes of an effective value-stream manager. Certainly the persons in charge of the value stream has to be senior enough in the company to command the respect of the team, and they have to be well versed enough in the broader aspects of the business to appreciate the importance of each function. They do not have to possess superpowers, however. They are team builders, not bosses.

To ensure a successful transition to value streams, each senior manager must make the single most important contribution of providing full, unqualified support to both the idea and to the value-stream managers. The greatest obstacle to value-stream success—and therefore to achieving the level of excellence companies such as Andersen and Parker Hannifin enjoy—is one or more senior managers who refuse to give up their turf. The vice presidenet or director of sales and marketing who argues eloquently and passionately why sales and marketing should not be included in the value streams but should continue to work for him while the rest of the company restructures can easily kill the entire transition to excellence. The vice president or director of operations who will support the value-stream idea only if the value-stream managers all come from her silo and refuses to allow operations people to work for anyone other than operations people stops excellence in its tracks. Those attitudes are all too common, and they doom the company to mediocrity.

Human resources clearly has an enormous role to play in the transition. Literally every job in the company will change, which presents an administrative challenge. The bigger challenge will be that of communicating what and why. The best approach is to inundate all employees with communications before, during, and after the transition, soliciting their input and support each step of the way. Converting a good-sized, traditionally functional-based company to value streams is akin to turning a battleship around in a river, and human resources has to be integrally involved, as well as supportive, at every turn to be sure nothing and no one fall through the cracks.

The task of actually breaking up the departments and assigning people to the value streams looks and feels a lot like choosing sides to play baseball in the sandlot. It is important to get the right balance, however, and to avoid having it become a personality contest. If one value-stream manager came from a supply chain background, for instance, then that supply chain can afford to have a less senior supply chain leader as the value-stream manager can pick up any slack. Another value stream whose leader came from engineering, for example, would require that a more capable supply chain leader join the team. The objective must be to assure that strength in every key area exists somewhere in the value-stream leadership team.

Product engineering can have a somewhat tricky role to play in determining what level of work to assign to value streams and what to keep out in a general R&D function. Product engineering runs the gamut from fairly straightforward product maintenance (i.e., minor tweaking for quality issues and to accommodate changes in the manufacturing process) all the way up to theoretical research. Somewhere in between is the appropriate cutoff, and, since the move up the scale from product maintenance to research parallels the scale from tactical to strategic, it is often prudent to start at the low end and move more product engineering into the value streams as it becomes the standard operating mode for the company.

Accounting and information technology (IT) are going to have an enormous amount of work to do in transitioning the company to value streams as a result of having to change how just about all of the company information is collected and summarized. How much depends on the nature of the company's information systems? Companies that rely heavily on database and spreadsheet applications tend to have a much easier time than those that are extensively committed to a complex ERP system. The changes in accounting and performance metrics will be substantial, however, and a major effort is inevitable.

Finally, this is perhaps the most significant change a company will make under the direction of a CEO. On one hand, an enormous number of details will have to be addressed before it is finished. It is not possible for the CEO to micromanage the transition to value streams. On the other hand, this is no time to stand on the sidelines and delegate. CEOs must be actively involved in the process and must make their presence known at every level of the organization—not to tell anyone the details of how to split up those two product configuration jobs into four part-time jobs but to assure that the people who are doing it know the importance of getting it done. The CEO has to be the philosophical leader and the force to keep the organization moving through myriad small obstacles—some real and some emotional. The employees in every organization that goes through the transition conclude about three months into it that value streams are the most senseless thing senior management ever did. A year after the transition, the people in the company think value streams are the best thing that ever happened, and they wonder why it took senior management so long to do it.

If this doesn't sound simple, consider the inverse case. Suppose the company were already aligned this way. It probably was back in its infancy when everyone wore multiple hats; then it morphed into the current complicated, functional mess. Then suppose a consultant came in and advised you to break everything up into disjointed, functional, physically separated groups. What questions and concerns you would have! How much time is wasted trying to coordinate and communicate among all of the disparate departments working to different agendas? How many meetings and e-mails does it take to get the people from one department to get something as simple as scheduling a meeting accomplished? How much time is wasted on office politics as people in different departments jockey for power, control, and prestige?

The big problem is that nobody in the whole company is responsible for profits and customer satisfaction. Only the CEO has that burden. Everyone else is working toward some departmental goal that may or may not result in overall success. How often does the company come up short in its profit goals yet have to reward individual employees with bonuses and raises because, even though the company didn't fare so well, some employees were "outstanding performers" because they hit all their objectives out of the park? This alone should be proof that things aren't hooked together so well in functional organizations.

No, at the end of the day value streams are very, very logical: a team of people with mutual clear-cut objectives that are the same as the CEO's. Functional organizations are the complicated ones; they are also inefficient and messy. It will certainly take some doing to convert to value streams, but, once in place, they represent the cleanest and easiest way to run any business.

11

Simple Accounting

Let's begin by making a few points very clear: generally accepted accounting principles (GAAP) may well be legally required, but they have no place in providing management with the information needed to succeed in a manufacturing environment. Period. End of that discussion.

Previously we cited the example of the expenses like the lights in the factory. Accountants use many different approaches to allocate the light bill to products—on the basis of payroll or labor hours, based on machine hours or floor space, by some complicated activity-based accounting scheme—but at the end of the day it really doesn't matter. Every method results in incorrect and misleading information. The root of the difficulty is that the lights have no relationship to any particular products, so however accounting goes about assigning some minute fraction of the light cost to any item it is wrong. Assigning a penny of lighting cost to a product may well be required by the IRS and others for the purpose of placing a value on inventory for the statements the company must provide to the IRS, but management should not for a second believe that the product actually incurred a penny in lighting costs. To take that penny and build customer prices on top of it, expecting the customer to pay for an allocated penny for this product, but not that one, is silly. To include that penny in the cost of the product when making a make-versus-buy decision is sillier. That penny has nothing to do with that product or any other product. It is simply a small part of the real cost of lighting part of the factory. Beyond that it is no more than the output of someone's clever, but meaningless, arithmetic.

If the allocated costs were only a matter of a few pennies, this whole chapter would be moot, but it is a lot more than a few pennies. In most companies allocated or assigned costs are 5 to 10 times greater than labor costs. It is not unusual to see a dollar of product costs that is made up of

60¢ in direct material, 6¢ in direct labor, and 34¢ in allocated mush. A third or more of the "cost" of the product is not real; it is accounting math driven by GAAP rules. With this menagerie of all sorts of costs driven by all sorts of things built into the cost of any individual product, the futility of calculating product profitability becomes apparent.

Of course, the cost of the lighting and the rest of the allocated expenses has to be covered somewhere, by some products, or the business will not make money. That does not mean, however, that every product on every transaction has to cover a proportionate amount of the allocated mush. Walking away from a sale if a customer is unwilling to pay for some amount that was arbitrarily allocated to a specific product is senseless.

The problem lies in the failure to look at the business, the product offerings, and customers holistically. We want to break every cost down to its smallest increment and assign every penny to something—the same idea as breaking the business down into silos and individual machines. We would like to think that we can calculate a whole cost for each product for each sale and can compare it with the price paid. If we can be profitable on each transaction, then the sum of all transactions will result in a profitable business. It sounds good in theory but fails miserably in practice.

Since there are no rules or laws dictating what management can or cannot do when it comes to internal, managerial accounting, it is up to the senior management team to make some decisions about how it wants to run the company. Ignoring the question or simply accepting the old standard costs because everyone is comfortable with them is a decision to knowingly allow yourself to be led down a path to destruction. Obviously, that is not an acceptable course of action for a team with a sincere interest in taking the company to higher levels of performance.

The starting point is to decide what it is the company wants the accounting system to do. The most important objective has to be clearly communicated, understandable information. Jerry Solomon,[23] vice president of operations at MarquipWardUnited, led the transformation of the accounting system there. He said:

> Traditional accounting is a foreign language to the folks in the shop who have to use the information to improve. How can they improve something

---

[23] http://www.isosupport.com/newsletters/articles/2006_Feb_MarquipWardUnited_Pkg_Lean%20 Accounting.pdf; Solomon, J. *Keeping Score With Lean Cost Accounting Management*, SME; February, 2006.

if they don't understand it? We have to make accounting information easy to understand, actionable, and timely. If you go to a football game, you always know the score, how much time is left in the game, and what yard line you're on. Imagine if you went to a game and you didn't know any of those things. It would be pretty frustrating. In comparison, that's what traditional accounting provides the folks who have to use the information. They don't know the score; they don't know where they are; and they don't clearly understand the goals. Yet we give them numbers at the end of the month and tell them to make them better.

He was referring to the accounting gyrations of amortizing, allocating, and assigning—all of which are other names for taking an expense and putting it somewhere or in some time period other than the one in which it really happened. A typical monthly profits and losses (P&L) statement contains very little information regarding the month in question, thanks to those GAAP-driven numbers games and something called the "matching principle" that drives the company to take this month's spending and bury it in inventory until the day comes that the products are sold. This month's cost of goods sold is composed of money spent this month, last month, the month before that—maybe even some spent last year. The first step, if management shares Solomon's objective of making accounting information "easy to understand, actionable, and timely," is to put as much of the accounting on a cash basis as possible. Let the purchased material portion of spending wash through inventory, but call everything else a "period cost"—one that is charged in the month, or period, in which it was actually spent. When managers, or more importantly the operations people responsible, see a $10,000 expense for machine maintenance last month, that has to reflect the real money spent to maintain machines last month.

Next, both revenues and expenses should be accumulated by value streams, not the old functional departments. The objective is not to run the lowest cost-machining department or the lowest cost-assembly operation. The goal is to operate at the lowest total cost. When expenses are broken down into chunks called "cost centers" or, worse, "profit centers" the ability to see and optimize the whole diminishes. It gets harder to see and implement an idea that spends an extra dollar in one part of the process to save two dollars in another part.

Next, the goal is not only to give people a clearer visibility of costs but also to have them take that clear vision and reduce the costs, especially the non-value-adding costs. In addition, most of that allocated mush is non-value adding, which makes it all the more important to provide visibility to the folks tasked with eliminating it. The best way to accomplish this is to assume that all costs, except for direct material, are fixed. It is often this step that creates the most discomfort for accounting and senior management, but it really is the most logical assumption to make.

All costs are fixed in the very short-term. As you read this there is not much of anything you can do to change the spending in the factory. For the next few hours or days the costs are fixed. At the other extreme, over the course of the next three years, everything can be changed, including replacing the entire senior management team and moving the entire business across the country or to China. So for the short-term everything is fixed, and in the long-term everything is variable; in addition, most decisions are made somewhere in between the two extremes where there are apt to be elements of both fixed and variable cost behavior and control.

What makes it more complicated is that just about everything in any given account is actually a combination of things. The charges to the machine maintenance account may include costs for repairs as well as costs for preventive work. They may also include the paycheck for the maintenance person who does both scheduled and unscheduled work. The breakdowns may be somewhat variable, whereas the scheduled maintenance and the person's labor are more fixed. Most accounts include these combinations. For years the management accounting experts have tinkered with more and more arcane variations of statistical tools such as regression analysis to try to figure out how to determine fixed versus variable costs with greater precision, all for naught. The answer to the question is unknown and unknowable, and it all depends on the time frame anyway; however, the question is not only unsolvable but also irrelevant.

Again, it boils down to what management wants to accomplish with the accounting system. If the goal is to continually reduce costs—or at least to keep them from increasing—then how does that play out if you make an error in calling costs fixed or variable?

SCENARIO #1

You use a standard cost system that basically assumes that all costs are variable; after all, once someone has gone through the exercise of allocating that penny for lighting costs to the product, then one penny is allowed to be spent for each one of those products made or sold.

As the business grows the accounting system allows additional spending proportionate to the increase in sales volume. So if a cost is really fixed but management has erred and called it variable, if the spending rate is $10,000 per month, and if volume goes up by 5%, the accounting system has allowed a fixed cost to be overspent by $500 per month without being flagged or highlighted as a spending problem.

In a steadily growing business, making the mistake of erring on the variable side takes away insight and pressure to keep fixed costs under control. They are allowed to creep up and essentially become variable even when they shouldn't.

SCENARIO #2

The accounting system assumes everything is fixed, even though some costs very probably have some degree of variability in their behavior. Now volume increases by 5%, but no increase in spending is allowed anywhere except for the direct materials that went into the products. Any and every account that went up by any amount jumps out as a spending problem, requiring investigation and explanation.

Clearly this approach puts a much greater degree of cost control pressure on the entire organization. This does not mean that some costs will not increase with volumes. After all, some costs really are variable at least to some degree, but they will go up only with management knowledge and clear visibility.

In reality, most businesses do not have wild swings in short-term volume. The variations from month to month tend to be small—less than 10% and often only in the 2–3% range. As a result, the operating mantra becomes to spend as much as last month or less in every cost category. This puts tremendous pressure on the entire organization to keep constantly engaged in costs and to be continually looking for opportunities to reduce them.

In the companies that manage with this approach to accounting (e.g., Buck Knives, Boeing, Parker Hannifin and hundreds of other companies implementing Lean accounting) they almost always have a volume "trigger" to cause them to go back and recalculate things if volume does change by a substantial amount. Normally the trigger is in the 10–15% range. If volume changes in the short-term by an amount greater than the trigger, these companies sit down and reestablish the monthly fixed-budget amounts.

So the key elements of the internal accounting systems must include (1) accounting for everything (except for direct materials and capital investments) as period costs; (2) accumulate, track, and manage revenues and expenses by value stream rather than any functional areas; and (3) call everything fixed except for the direct materials. What you end up with is something that looks a lot like a manager's home budget. More importantly, you have a set of accounting reports that say what they mean and mean what they say. They will be very easily understandable by everyone in the organization. They will meet the MarquipWardUnited aims of being easy to understand, actionable, and timely.

Standard costs are by the wayside, as the only specific product costs are the direct materials. This typically creates a great deal of consternation because, as inaccurate as they may have been, the old standard costs were the universal tool and the universal security blanket. Unfortunately, they also served as the buffer between the sales and marketing group and the operations people. In most companies the sales and marketing people have little knowledge of the operating costs because they do not need to. The standard cost tells them all they need to know. It is up to them to take those products at that standard cost and go sell them at a price that covers the selling, general, and administration (SG&A) expenses of the company and leaves a profit. If they cannot do so, they are off the hook because the standard costs are too high and that is not within their control.

On the other side of the standard cost buffer, operations people know very little about the pricing side of the business or the financial interactions with the customers. Their job is to make things at standard, and what happens from there is not their concern. When the standard costs go away, sales and marketing come face to face with operations, and they have to collectively figure out what to do. How deeply they have been isolated from the other side and how little the managers really know about the overall business often become painfully apparent.

Typically when presented with the idea of getting rid of the old GAAP-based standard costs, the people in accounting have little intellectual objection to the idea; after all, they are the ones who did all of the amortizing, assigning, and allocating, and they know better than anyone how far from reality the numbers are. Their concern is more often a very valid one, and that is the financial ignorance of the rest of management. Managers in standard cost-driven companies tend to be masters of the system (and often masters at gaming the system to hide problems and make things look better than they really are). Accounting is apt to have a legitimate concern with the ability of people in sales and marketing and operations to manage without the safety net of standard costing.

The biggest concern is usually in the area of pricing—an area so important it will merit its own chapter later in the book. Beyond pricing lies any number of decisions, ranging from new product costing that drove design decisions, make-versus-buy decisions, and capital investments that all had standard costing as a key input. The decisions made with the costs were not good ones because the standard cost data were not good, but the decisions tended to be safe ones. Standard costs may have kept the company from making sales that would have been helpful to profits, but they also assured that nothing was ever dangerously underpriced. The answer is that accounting must change from a number-crunching, report-generating role to a teaching, coaching, and supporting role. The organization will be given more information in the form of rawer data—accurate but unprocessed—in fact, accurate because it is unprocessed. Accounting will have to be a major player in the discussions and decisions that take that data and do something with them, at least until the overall financial skills of the company rise.

The final major area of concern is direct labor, which has traditionally been viewed as a classically direct cost. The assumption underlying calling direct labor a variable cost is the notion that people can be laid off and recalled with the changing winds of sales volume and that new people can always be hired to support growth. In Chapters 5 and 6 concerning culture, when the company embraced the commitment to provide security and value to employees, all of that came to an end. Once the excellent companies make such a commitment to their employees—that there will be no layoffs unless it is the last resort before insolvency—direct labor most certainly became a fixed cost. The expectation is that, in return for the company's commitment to its employees' long-term security and the value the company will strive to bring to their lives, they will actively engage in continuous productivity efforts. Fear of working themselves out of a job will be gone, and again within some reasonable upper limit, the employees will find ways to meet demand increases without the need to hire many additional people. As a result, direct labor should be largely fixed on the upper end as well.

For operations and human resources, this will have a profound impact on scheduling and overtime policies, and most importantly it will put a very high premium on cross-training. When pressed to continually improve productivity, the most effective tool will be having people who are capable of performing a broad number of jobs. This too will be addressed in a later chapter.

As a final cautionary note, there has been considerable change and evolution in management accounting thinking in the years since Tom Johnson and Robert Kaplan wrote their groundbreaking book, *Relevance Lost: The Rise and Fall of Management Accounting* (1991), which first outlined the more harmful effects of traditional accounting on manufacturing. The first solution was activity-based costing (ABC), which has since been largely discredited. It is really just an attempt to make allocations more accurate, which is an effort that could possibly succeed. Don't waste your time on ABC as a compromise solution. The work currently being done under the auspices of Lean accounting, as well as some of the work referred to as throughput accounting, is the only effective approach to bringing accounting to bear as a management tool for continuous improvement rather than an obstacle. That is largely because it is the simplest approach—and simple means useful and accurate.

12

The Roadmap from Chain to Channel

The value-stream map is a flowchart describing how value is created for the customer from start to finish—from the inbound supply chain to the outbound distribution channel. It describes the core of the business and is the single most important document management has. Every piece of information managers have been using to run things is nothing more than a bit of data along a value-stream map. Knowing labor efficiency or the cost of something is neither good nor bad unless you can see how it fits into the big picture.

While most of the Lean consultants and Six Sigma wizards will have you use value-stream mapping on a project basis, in the excellent companies it is a living document, visually and vividly displayed and continually updated with information that enables everyone in the value stream to see how things are going, where any problems exist.

The core of a value-stream map is a sequential series of boxes identifying each step in the overall process. Of course, any number of "feeder processes" flow into the main line, indicating everything from subassemblies to business processes to schedule events, or to provide specifications or machine programs needed to execute customer orders. At each box, data are recorded concerning cost, number of people, the identification of constraints, quality results, and buffer inventories. There are no rules or limits to the types and quantities of data required to manage by the value-stream map. Contrary to many "Lean manufacturing experts," no special forms are required, and no specific data are mandatory. There are no rules mandating that inventory be placed in one particular-shaped identifier while computer operations are marked in another. All that matters is that the value-stream map provides a comprehensive visual picture of the entire process and that the information needed to optimize the process is clearly available.

While there are no rules when it comes to value-stream mapping, there are a few critical underlying principles, the most important of which is to link it from end to end. Just as when taking a trip, information about a specific location is useful only if your map shows your route from start to finish so you can see where you are along the way, a value-stream map is useful only if it encompasses the entire process—what Taiichi Ohno, the original Toyota manufacturing leader, defined as "the timeline from the moment the customer gives us an order to the point when we collect the cash." A value-stream map that does not link end to end with customers is only marginally better than no map at all.

It is also necessary that the map be a manual, visual document. The power of the value-stream map is that it provides a clear image of the entire process. Putting the value-stream map into a computer defeats the purpose. The Bible says, "Nor do men light a candle and put it under a bushel"; in the same way, it is foolishness to take the visual representation of the core business process and then to bury it in a computer so that only certain people can see it. Most of the excellent companies use an entire wall of the war room in which the value-stream management team works for the value-stream map. Most effective is the use of a floor-to-ceiling, wall-to-wall whiteboard, with the operations identified by pieces of paper taped to the board and plenty of space around each operation for whiteboard notes and continually updated information. In one Wahl Clipper value stream, the value-stream map is on an 8-foot whiteboard mounted on rollers so it can be moved around the factory to provide the necessary perspective when issues are addressed in one work area or another.

The value-stream map is the vehicle that pulls all of the concepts previously discussed together, including quality, accounting, supply chain, metrics, and people.

When we discuss quality management in Chapter 15 we will go back to the value-stream map and will introduce the idea of defect mapping on the same simple visual control. The defect map is a step-by-step identification of each opportunity to create a defect. For instance, at a machining operation, each critical tolerance that must be met represents a defect opportunity. The steel itself also represents one or more defect opportunities—the dimensions, grade, and hardness of the steel being machined, for instance. The value-stream map should identify each such opportunity to do something wrong (i.e., to miss a spec) and the measures to preclude it from turning into a defect that reaches a downstream operation or worse—one that finds its way into the hands of the customer. Finally, the value-stream map is the place to track the results of the defect map, indicating how well the defect opportunities are avoided and handled.

The value-stream map is also used to indicate where along the process expenses are incurred, including support labor and other overhead costs. By providing visual representation it is easier to see where non-value-adding costs are being added. The notation of such costs highlights where value-stream management attention can pay the biggest dividends in optimizing their Value Added Ratio (VAR) calculated by dividing value adding cost by total costs. Typically, the location of material handling, maintenance, and quality inspection personnel are noted in a unique color providing a visual image of waste. Also, the span of responsibility of supervisory people is marked, making it readily apparent where the value-adding people require too much oversight.

Value-stream mapping is not something to be used as an adjunct to whatever complicated computer system you might be using, replete with dashboards and all sorts of bells and whistles aimed at setting senior management up at the business equivalent of NASA's flight control center. It replaces most of that approach. It is the antithesis of that approach. It is a basic picture of how work gets done, and you use it to plan, manage, and improve the core of the business in a manner everyone in the company can see and understand.

13

Scheduling the Factory

Nowhere is the difference between wasteful, ineffective complexity and cheap, effective simplicity more evident than in how companies schedule production.

Excellent companies use whiteboards, spreadsheets, and *kanban* cards to continually schedule the factory. Mediocre ones spend absurd sums of money on ERP/manufacturing resource planning (MRP) systems and deploy an army of planners and schedulers.

Excellent companies put enormous energy into continually reducing cycle times to get their lead times down and responsiveness up. In the mediocre companies supply chain people gripe nonstop about the lack of accurate forecasting, and salespeople complain day and night about needing more inventory.

In the excellent companies most of the material is ordered from the suppliers directly from the shop floor with a production person firing off an e-mail or making a phone call to the supplier. In the mediocre ones a designated buyer makes a bunch of computer transactions and analyzes part requirements and on-hand quantities before killing a few trees to create a purchase order.

In the excellent companies scheduling problems are assumed to be the result of unnecessary complexity, and solutions are aimed at cutting to the core of the shop floor decision-making needs and empowering the production folks to be self-scheduling. In the mediocre ones, scheduling problems are assumed to be the product of bad attitudes or a lack of intelligence on the shop floor for their failure to appreciate and comply with the beauty and grandeur of the mainframe computer-generated scheduling results—so the solution is more computer-centered complexity.

While scheduling does not have to be nearly so difficult as we make it out to be, it does require some knowledge and a solid grasp of some core principles. It is more important that members of senior management understand those principles than having the people actually doing the scheduling know them. The first of them is the basic water-and-rocks analogy.

Consider production to be like the on-time delivery canoe cruising down the value-stream river through the factory. The river (factory) has all sorts of problems—bad layout, poor quality, unreliable machines, poor scheduling resulting from long lead times, unmanaged constraints, and who knows what else—but all of them are non-value adding and are driving you to have to charge higher prices for stuff customers perceive to be creating no value. Those problems are like rocks in the river. Inventory is the water in the river. If you put in enough water, the canoe can coast easily over all of the rocks and reach its destination, the customer.

Now managers have two choices. They can keep all of those rocks covered up with inventory all the time so they seem to be invisible. Just because they are covered, however, doesn't mean they aren't there, eroding profits and undermining the finished quality. The inventory just makes sure the rocks aren't causing any problems today. Or managers can continually lower the water level in the river, bringing more rocks to the surface and forcing the plant to address the problem: eliminate the cause, and enable the canoe to coast to the customers. The first choice—drown the factory with inventory—is the easy one. It is also the more expensive one. Not only do the costly problems stay hidden, but all of that inventory costs money too. It gobbles up floor space, requires someone to move it around and cycle count it, often incurs property tax, and so forth. Coupled with accounting systems that allocate and bury all of these overhead costs, the company has little chance of success. The second option—drain the swamp and bring the problems into the spotlight so they have to be addressed and solved—is what the Lean manufacturing folks are talking about when they advocate continuous improvement.

While the accounting folks call inventory an "asset," in a manufacturing organization inventory is only an asset in the same manner as sand is an asset to an ostrich. That is, it is very helpful if your objective is to delude yourself into thinking everything is fine. However, that is about as far from the dictionary definition of an asset—"a useful or valuable thing;

an advantage or resource"[24]—as you can get if your objective is to reduce costs and improve cash flow.

Principle number two is that inventory and time are two sides of the same coin. Time is inventory, and inventory is time. If you have 1,000 widgets in inventory and you sell 50 a day, then you have 20 days' worth of widgets. Straightforward enough. It can also be said that assuming you do a pretty good job at rotating the stock and practicing first in–first out if you buy another widget today and add it to the pile, it will be 21 days until that widget gets shipped out.

In the flow from the supply chain through your business and out through the distribution channel to the final customer, the length of time material will spend in your place, consuming fixed costs and covering up your rocks depends on how much material you have. More inventory means longer cycle times. When the folks at Motorola said, "The best quality producer is the shortest cycle time producer, and the shortest cycle time producer is the best cost producer," it was saying the company that lowers the water level and exposes the most rocks will have the best overall cost structure because it did the best job of rooting out all of the hidden cost in the factory.

When Taiichi Ohno at Toyota said, "All we are doing is looking at the timeline, from the moment the customer gives us an order to the point when we collect the cash. And we are reducing that timeline by removing the non-value-added wastes," he was talking about the same thing.

When Henry Ford put everything in motion and figured out how to build cars in a week or so from start to finish, he did so by getting rid of even the smallest pebble in the Ford factory river. He rooted out just about every possible penny of non-value-adding cost and was able to sell Model Ts at absurdly low prices and still make a lot of money. Put another way, he created a fantastic value proposition for his customers and himself. With minuscule non-value-adding costs he was not in the position of having to convince customers to pay more for cars than they were worth, and he did not have to see his profits diminished by a lot of money spent for no useful purpose.

The strategies of Lean manufacturing and Six Sigma have been convoluted to mean a lot of things other than compressing cycle times to expose the rocks and force the elimination of costs that do not add value for the simple reason that accounting is built around the idea that inventory is an

---

[24] http://www.yourdictionary.com/asset

asset—just as valuable as cash. When senior managers don't understand that inventory may be an accounting asset, but a manufacturing crutch and the device that covers up waste, they see little value in compressing cycle times. Sales folks who advocate more inventory—"You can't sell from an empty wagon" and other clever criticisms of manufacturing—are really advocating more water to cover up the waste that is causing them to have to convince the customers to pay higher prices.

Cycle-time compression, Just In Time, and running with Leaner inventories are not about inventory, per se. Nobody cares about inventory for inventory's sake. It is about continually bringing problems to the surface and forcing resolution, which continually pushes the factory to less non-value-adding waste.

Senior management needs to clearly understand these principles to give direction and support to the scheduling function. Make no mistake: scheduling is very, very easy. It is managing all of the excess inventory that makes it hard. If it weren't for all of the underlying problems that drive the factory to carry so much inventory, all a scheduler would have to do is write the order down on a piece of paper when it comes in and hand it to the factory, telling them to "make this next."

Inventory has another big driver: supplier lead times. When your customers want you to ship in one week but your suppliers can't ship parts to you for four weeks, you have a problem: a three-week gap. You have to be ordering parts four weeks ahead of time, but you know what you actually need only one week ahead. So schedulers try to solve this problem by (1) pounding on salespeople to predict the future out to the lead time with perfect accuracy (in effect, asking them to predict orders with such accuracy that to the schedulers they are getting four-week customer lead times); and (2) bringing in an excessive amount of purchased inventory to try to cover any eventuality. Neither approach works too well, so the scheduling job becomes a grand exercise in managing parts availability and shortages. There is nothing simple about scheduling in an environment in which supplier lead times are way out of line with customer delivery terms.

Of course, the problem in many manufacturers is a whole lot worse than figuring out how to cover a three-week gap. That would be a walk in the park for most production schedulers. The reality is that, because accountants and senior managers are oblivious to most of this, getting parts from some cheap source 9,000 miles away looks like good economics, and lead times from China are measured in months rather than weeks. When the gap between supplier lead times and customer lead times is stretched out to 10 or 12 weeks, things really get interesting. The problem is so great that the two solutions (i.e., forecasting and carrying big inventories) become epic in size. Companies spend tens of thousands of dollars on forecasting software with ever more sophisticated prediction algorithms and create jobs like *forecast analyst* and *demand planner* and hold forecast meetings to try to sort through it all to predict the unpredictable. The investment in forecasting is trivial compared with the investment in ERP software, the basic purpose of which is to sort through all of the inventory—a veritable tsunami in the factory river—to keep production going. All of this is necessary to achieve the savings from buying parts in China or some other faraway place. Those savings become very expensive indeed. The vast inventory needed to cover the long lead-time gaps, added to the big work-in-process inventories to cover up all of the factory rocks, results in big scheduling departments and enormous factory complexity.

The key to excellence is merely stepping back from the whole mess and realizing that, rather than throwing lots and lots of good money after bad to try to manage through all of the complexity, and simply reducing supplier lead times; voilá—problem solved. That is why Just In Time came to pass. How to go about accomplishing this will be addressed in the next chapter. The next order of business is driving the work-in-process inventory out of the factory—lowering the water level to force the waste out of manufacturing. This is what the rest of the Lean manufacturing and Six Sigma toolkits are intended to do.

Tools like set-up reductions, *kanbans*, and mistake proofing were designed to help the factory drive down the cycle times. Using them as tools to reduce direct labor, for instance, is like using a hammer to drive a screw; it's the wrong tool for the job and not likely to get the desired results.

There are two other key principles senior management needs to master to lead the factory scheduling function: push versus pull and some basic statistics.

The supply chain management people talk in terms of push and pull, which means nothing more than working to a forecast versus working to reality. We have described the process and the ensuing problems from trying to manage to a forecast. That is a "push system." It means pushing material and production into the factory and into inventory ahead of real demand. Pull, on the other hand, or *kanban*, means carrying a set level of inventory and continually replenishing it based on actual demand. It works a whole lot better. *Kanbans* can be set up anywhere there is repeat demand—in finished-goods inventories, works-in-process, or purchased parts. The process is to simply analyze how much inventory is required to protect the gap in the lead time and any variability in the flow and to put that quantity in place. Then when customer orders come along and consume some of the inventory, the scheduler issues the orders to replace it in kind. It is simple and assures that inventory is at least capped and will not increase based on forecast errors. The idea, then, is to identify why the *kanban* inventory is the size it is—long lead time and variation in factory flow—to improve those variables, and to reduce the size of the *kanban*. That is actually how the water-level reduction takes place. Put everything you can into a calculated *kanban*, do all the scheduling on a simple replenishment pull (which can be done on the back of a bar napkin; no big complicated computer systems needed), and focus your attention on reducing the size of the *kanban*.

The statistical principle at play in directing the scheduling function is pretty basic. The amount of inventory you need is a function of both the lead time and the potential variation in the rate of demand. That means the calculation has to be a little more complicated than having someone gaze into a crystal ball and say, "Let's keep four weeks of inventory on hand." That usually means four weeks of average demand, and averages are horrible numbers to use. As the old saying goes, "If you stick your head in the freezer and your feet in the oven, on average you should feel pretty good." You have to use any one of a number of formulas available that takes into consideration both the number of weeks of lead time you have to protect and how much variation there is apt to be in the customer orders during that time.[25]

Most importantly, senior management has to engage in the scheduling process. In far too many companies, rather than viewing the factory scheduling process as a critical driver of the strategy to drive out waste, it is seen as some sort of clerical exercise requiring relatively unskilled people to process paperwork in accordance with the dictates of some anonymous computer system.

---

[25] There are a number of different statistical formulas for calculating *kanban* quantities. The formula used by author Bill Waddell in his consulting practice is *kanban* quantity = (Average weekly demand + (2Σ * 1.1)) * ((Cumulative Lead Time – Customer Lead Time)/2)).

14

Purchasing

Unless you are very vertically integrated, the purchased content of your product represents: most of its cost, is the overwhelming driver of its quality, consumes the greatest amount of your cash in inventories, and pretty much controls whether you make a profit. In fact, it is the single biggest driver of whether you will grow and prosper or go under. So the logical thing is to do your buying via an online auction and turn it over sight unseen to the lowest bidder, or source the key components with some factory in China you have never seen via some agent you found on the Internet, right? Or maybe just give it to the sales rep who drops in and takes your buyers to lunch and springs for a bottle of scotch at Christmas. Put that way, the approach most manufacturers take to the sourcing and purchasing function sounds outrageous.

Engineering, in particular, has to be fully engaged in viewing the sourcing function as a very strategic endeavor. Your product cost is as much as 90% fixed when it leaves design. All the buyers and shop-floor manufacturing people can do after that is tweak it up or down a little, but the cost structure is a done deal before they even see it. Engineers who design around parts from a catalog without considering the supplier's lead times and quality controls, are a manufacturing nightmare. Adding suppliers who sell only one or two parts to you expand your non-value-adding cost exponentially. Engineers who assume they know all there is to know about the materials and components available usually leave a lot of potential value on the table. The right relationship with the right suppliers means you can avail yourself of a lot of free technical input and a mutually, economically beneficial setup for both parties. Engineering team members have to be a fundamental part of the sourcing process, but they have to fully understand the profit objectives of the business and know how to optimize the whole. They cannot view product design as something akin to the junior high school science fair where the only thing that matters is technical elegance. Design engineering is, when you boil it all down, the technical link between the supply chain and the final customers out at the very end of the distribution channels.

Accounting, of course, has to either get involved in assisting with a broad, strategic understanding of sourcing or get out of the way. In the value-stream section we described the inventory out of control with a root cause of design engineers who had been driven to explode the variations on a pretty common part because all they knew to consider was the purchase price of the parts. A hundred variations on the same part from a dozen different suppliers makes sense only if you apply typical accounting logic rather than basic common sense. The least educated shopper knows that you can spend all day running around to six different grocery stores to do your weekly shopping to get the lowest prices in town on each item you buy, but the $5 you will save will be more than gobbled up by the $8 you will spend on gas, let alone the waste of an entire day. So you go to the one or two that enable you to optimize the whole. The purchasing function in your business has to be executed with the same simple approach. Like the shopper who settles in on the one grocery store that, all things considered, represents the best overall selection and value, you have to look over the landscape and pick a couple of suppliers that are the best situated to support a broad range of your needs.

That broad range of needs includes technical, economic, and strategic considerations. Technically, you have to develop relationships with suppliers that can share your vision of growing in the distribution channels you serve and are willing to provide input to your product development efforts to make sure you are always offering the best value. Economically, your suppliers have to share your understanding of value. We previously described supplier partnerships as "characterized by mutual cooperation and responsibility, as for the achievement of a specified goal." That goal is continually increasing the value you jointly provide to customers. Haggling with suppliers over pennies that will either come out of their profits or yours, depending on who has the superior negotiating skills, is an utter waste of time.

You have some value content in your overall cost structure—say, 60% of the money you spend goes to creating value, and the other 40% goes to non-value-adding wastes. Your supplier also has some percentage of value to waste. The pricing and sales you realize from manufacturing are a function of the 60% you spend on things customers will pay for. There you have the "specified goal" you and your supplier should mutually cooperate upon: improving your value ratio. To the extent the supplier can help you spend more money on value and less on waste—from defects, inventory costs, paperwork, and the like—you should be willing to pay more for the things you buy from that supplier. At the same time, the supplier should be working with you to enable you to improve its value proposition—eliminate its waste. The mutual economic gains cannot come from arm's length negotiations in which each of you is keeping your cards close, looking to play a zero-sum gain; your gain must be the supplier's loss and vice versa. In the end, you both lose that game. Its only possible outcome is an eventual end when one of you says enough and you both go off and find another entity to begin the process all over again.

The final consideration is strategic. We discussed values, supply chain strategies, and the importance of setting your factory up as Peter Drucker's "wide spot" in the stream. You have to make senior management-level decisions about potential suppliers to evaluate all of the nonfinancial intangibles and whether the supplier shares your values and long-term goals. This does not mean that the leadership team necessarily needs to be involved in every supplier selection. It does mean, however, that a well-thought-out policy should be established by the senior team identifying the broad criteria to be used in selecting suppliers. The selection of suppliers of critical technologies and high-cost components should have direct involvement of the senior team.

The role of purchasing should be primarily one of facilitating long-term, broad relationships and not to serve as a constraint. Once the relationship is in place, purchasing's role should be to act as the matchmaker. Bring your accounts payable people into direct contact with the supplier's accounts receivable folks; introduce your receiving dock personnel to its shipping people; get your engineers and suppliers into direct communication; and then get out of the way and let the relationship work at every level. Demanding that all communications go through some buyer is a wasteful formula for failure.

More important than the price for each item are the terms between you and the supplier for overall supply chain management. Once you have come to understand that reducing supplier lead times is the key to enabling you to root out all of the destructive waste in your factory, the need to do something better than to throw purchase orders at your supplier one at a time as you need parts becomes more obvious. The simplest way to reduce supplier lead times is just to ask the supplier to carry finished-goods inventory. This, however, is often no more than shoving the inventory risk and carrying costs back on the supplier and doesn't really help the overall supply chain. And it serves to cover over the waste in the supplier factory, which will eventually be passed on to you.

The use of blanket orders is almost always more effective. A typical example would be to estimate your use of the item for the next year, then give the supplier a blanket order for about 60–70% of that quantity, and then ask it to carry inventory to support that volume. You tell the supplier what you actually expect to use and guarantee about two thirds of that amount. That way the supplier knows its investment in the inventory is not at risk. If all goes well, seven or eight months into the year you have bought all the blanket-order quantity, and you do the same thing all over again. In the meantime, the supplier carries inventory sufficient to meet your needs, while reducing its internal lead times.

Often the best way to start is to sit down with the supplier and find out why its lead times are so long. More often than not, it has to do with its lead times for raw materials. If that's the case, you might be buying a dozen different products from the supplier that use the hard-to-get material, and the problem is that you don't know which of the final configurations you will need. You do know that you are going to use something, however. So you either guarantee some amount of volume in total without specifying the particular part numbers, or you even buy the raw material yourself and keep it at the supplier's site. The point is that by investing in $1 of raw material for the supplier—or guaranteeing the supplier's investment in it—you can save $5 worth of inventory in your factory, and achieve a substantial reduction in lead times.

It usually takes nothing more than a little creativity and hard work to reduce supplier lead times by big amounts, but it will never happen if you don't try. And it certainly won't happen if you never ask the simple questions of your suppliers and then recognize that the risk of inventory should be equitably shared, as should the return from that risk when you turn shorter lead times into better value for your customers.

15

Quality Management

Any husband who has ever felt the pressure of moving a large piece of nice furniture through a narrow doorway under the watchful eye of a wife admonishing him to be sure not to scratch either the furniture or the door frame understands the basic concept of Six Sigma only too well. The originators of the concept at Motorola back in their heady, Baldrige Award-winning days when Six Sigma burst onto the manufacturing scene used a similar analogy—that of a car and a garage door. Either way, the wider the door, the less likely you are to fail. If you try to drive a 7-foot-wide car through an 8-foot-wide garage door or you try to move a 30-inch-wide armoire through a 32-inch-wide door, there is not much room to stray from the centerline before you hit something and find yourself in a lot of trouble.

In work terms, the width of the door represents the capability of the process. The question is how much room is there for something to stray from perfection before it becomes a real problem? Obviously the wider the garage door, the less likely you are to hit something. Basically, Six Sigma theory says you need to have a 14-foot-wide garage door to operate with confidence that your 7-foot-wide car will never hit anything (at least as close to never as never can be in the real world).

Motorola's big idea was that processes tend to drift out of control—away from perfection—and that perfect quality is possible only when you plan for this and assure that every process is robust enough to tolerate a great deal of drifting without causing a defect. As we mentioned previously, the broader intent was to assure flow. Variation in the rate of flow through the factory gums up everything. Only when each step in the process receives a steady input from the previous step can things happen at their lowest cost. When production is fed to the next operation in fits and starts, machine and people time are wasted, inventories build, and non-value-adding costs start to add up. So by making each step capable of handling a wide margin of error—especially human error—the flow will be able to continue, and the factory will operate at its best. Put another way, the factory is a series of dozens, perhaps hundreds, of garage doors. The presence of just one or two doors without enough room for steering error will stop the flow of cars driving through the factory, creating backups, logjams, inventory, and excess cost.

That is Six Sigma in a nutshell—at least Six Sigma as it was originally conceived. Since then it has strayed quite a bit and has become shrouded in special terminology, procedures, and forms, and it requires people certified to hold belts of varying colors. It is often applied as some sort of grand cost-cutting technique rather than as the steady-flow driver it was supposed to be. You will notice, however, that a very small percentage of Six Sigma cost savings finds its way to the bottom line, and General Electric under Jack Welch, the grandest of Six Sigma users, generally used it to identify outsourcing opportunities. This is not the sort of Six Sigma any of the truly sustaining excellent manufacturing companies deploy. Theirs is truer to the original concept, and they rarely require Six Sigma black belts to put the principles to work. It really just comes down to management integrity: whether management can be honest with itself. If you want Six Sigma to be a self-fulfilling prophecy and have it tell you what you want to hear—that outsourcing is the answer to everything—you can manipulate it that way. If you want to use it as a tool to get to the truth of things, however, it is a very good tool.

The ideas about process capability, process design, and process control are really the power of Six Sigma thinking. They have a lot in common with the tried and true statistical process control (SPC) applications. SPC essentially says that every process has its center point—perfection—and a tolerance limit. The tolerance limit is the point at which the process has strayed so far from perfection that it no longer works. SPC charts track the straying and enable someone to take action before defects are produced.

The idea of process is central to achieving outstanding, near-perfect quality. In poorly run companies, a defect is cause for management to ask, "Who screwed up?" In well-run companies, the question is, "What process is not under control?" And what do we need to do to make the process robust enough, or capable enough, so as not to allow defects to be created? Anytime quality is dependent on people never making mistakes, quality is going to be pretty shabby. The idea behind the design of a good quality system is to expect human errors and have the processes capable of absorbing them without creating poor products.

In designing the quality system, the senior management team has to be composed of process thinkers. That should be sounding like a broken record by now. Organizing into value streams is nothing more than formally organizing people around processes rather than functions. In managing constraints we are simply identifying the greatest obstacles to process flows and paying very close attention to them. In overhauling the accounting systems we are primarily trying to organize financial information and decision making around process performance. That recurring theme—process, process, process—applies to quality as well. Effective quality management requires creating and controlling processes capable of (1) preventing quality problems from occurring in the first place, and (2) quickly catching them when they do occur.

Manufacturers in a constant struggle with quality are those that are siloed and compartmentalized. They think that quality is something the quality control department and operations handles with maybe a little engineering support and that poor quality is a failure on the part of those departments. Sales and marketing, supply chain, accounting people, the CEO, and IT folks set themselves up as critics or take kibitzing to high levels to demonstrate their support but do not see it as their responsibility in any way. In fact, excellent quality is the product of an integrated management system, and the people in those functional roles have a very direct impact on quality.

In many companies, the toughest quality problems are those revolving around vague standards for cosmetic features initiated by sales and marketing people. Hitting a ten thousandths of an inch spec on a machined component is easy compared with coming up with exactly the right shade of purple, the right feel on a gel grip, or perfect clarity on a blister pack. Setting measurable standards and then devising processes that can meet them are the fundamental building blocks of quality and require the active participation of everyone to see that this happens.

More importantly, sales and marketing reps must assure that customer quality requirements really reflect the customer's true perception of value. The sad fact is that very often the difficult cosmetic quality requirements the factory struggles to maintain are not about customer value at all but are driven by an attempt to create the illusion of value. The customer really wants a device that works well and holds up over time in hard use. Because the company has so gutted the core of the product in cost-cutting efforts, marketing has had to resort to wrapping the product in a purple gel grip in a crystal-clear blister pack to create an impression of quality that does not really exist. Quality is all about knowing what is truly important in the eyes of the customer and then getting those things absolutely right.

The technical quality battle driving every aspect of the product that make it work right and stand up to use is largely won or lost in the engineering effort to design products. Whether a high-quality product will be made and shipped is decided before it ever hits the factory floor. We do not want to turn this into a technical–statistical dissertation, but it is important to know that the ability to make quality products is largely driven by the standards, or specifications, established by the design engineers relative to the capabilities of the people and machines that will make them. Part of the Six Sigma body of knowledge and techniques is an analysis called *Cpk*, or a *capability analysis*. This is where that width of the car compared with the width of the garage door arithmetic has to be done. If the product specifications call for a tolerance of plus or minus eight thousandths of an inch and the machine that will make the part is capable of holding tolerances of only ten thousandths of an inch, it is theoretically possible to make the part, but there is not much room for things to stray from perfection on the machine before there is a problem. Before any new product is released for production, sales and marketing, operations, and engineering should review a detailed Cpk analysis of the product and have a high degree of confidence that the product is really manufacturable.

The design of the quality system is important to the entire management team for a couple of reasons. For one, it must be good. If the company is going to be focused on value to the customer and pricing is going to be value driven, then assured quality has to be a given. Value in the eyes of the customers does not end with very high quality, but it most certainly begins there. No company is ever going to reach excellent status if its quality levels are unreliable.

For another reason, there are a lot of bad ideas out there on how to create and control quality, and it is very possible to spend a lot of money and effort on quality without getting much in return. Just as many companies get caught up in the trappings of Six Sigma and fail to grasp the basic, underlying principles that make it worthwhile; it is very easy to substitute

International Standards Organization (ISO) procedures for ISO intent. Creating and filling out forms and putting together and maintaining the policy manuals needed for ISO registration is a waste of money. For some companies it may be a necessary waste of money as certain customers have made ISO registration mandatory, but it is nonetheless waste. The principles underlying the ISO process are what create value. It is so important to keep focused on that. Far too often, the senior management team is disengaged from the true quality system, assuming all is well because people are maintaining the ISO documentation and because some outside entity comes in every now and then and audits things. That is not the same thing as having an effective quality system. It means only that people are complying with a set of procedures that may, or may not, be getting the job done. And even if they are getting the job done, they may be drowning quality control with unnecessary paperwork and procedures.

Most important is the relationship between quality and flow. When one of the authors learned Six Sigma soon after its origins, one of his most influential teachers was Motorola's director of materials management. While continuous, defect-free flow across the factory results in the best cost, continuous flow also enables the best quality. They feed on each other. A Motorola Six Sigma mantra was, "The best quality producer is the shortest cycle time producer, and the shortest cycle time producer is the best cost producer." This relationship among quality, cost, and cycle time (how fast something goes from the beginning of the manufacturing process to completion) is a core principle that has been lost as the certified black belts try to apply Six Sigma to traditional cost cutting. It boils down to understanding that inventory undermines quality.

We cannot stress enough the importance of a keen appreciation of time to achieve manufacturing excellence. Whoever first coined the phrase, "Time is money," was a genius. Inventory is a lot of things, but most importantly it reflects time. The excellent companies continually drive down inventories to save time—not to free up cash or to shove all of the inventory risk and handling costs back on suppliers but because inventory consumes precious time.

If assembly needs to consume 10 widgets per day to meet customer demand and the factory keeps a buffer inventory of 100 widgets in front of assembly to be sure it always has plenty to work with, then the widgets being made today in machining to go into that buffer inventory will not be used in assembly for 10 days. If the widgets being machined are defective, no one will know for 10 days until assembly goes to use them and finds that they don't fit. The more inventory there is, the longer the time gap between creating a defect and finding it. Of course, the longer the time gap, the less likely it is that anyone will ever find out what went wrong. Often it is impossible to learn who made the defective parts or on what machine since the problem occurred weeks ago.

It gets even worse because the usual reason for having the inventory is to enable the machines making the parts to save labor costs by building big batches without having to lose time setting up the machine so often to make something else. As a result, if the machine making the defective widgets makes them in batches of 50 at a time, when the defect is found 10 days later, it is not one bad widget—it is 50.

Excessive time between creating quality problems and finding them makes it more difficult, if not impossible, to ever have excellent quality. Time between something going wrong and finding out about it increases the magnitude of the problem. Inventory is the manifestation of time. The more inventory there is, the worse quality is, and the more expensive poor quality is.

How the supply chain is constructed has a huge bearing on quality. These basic principles do not apply just to internal inventories and defects between operations in the factory; they apply everywhere. Purchasing in big quantities to save a few cents on the purchase price increases the likelihood and consequences of supplier quality problems. When salespeople complain that they "can't sell from an empty wagon" and that the company should carry more inventory, they are upping the probability of shipping poor quality products to customers.

Everyone on the senior management team has a stake in minimizing inventories and, as a result, shortening the cycle time between events in the process. This time–quality relationship is not just a manufacturing issue either. When people process paperwork or enter data into computers in batches, the same risks increase. The longer the time gap between when something is done and the person using that "something" gets it, the greater the chances and impact of mistakes.

Companies try to deal with this by having inspectors looking at the widgets to make sure no bad ones go over to the stockpile in front of assembly. Inspection is an utter non-value-adding waste of money and is the least effective way to catch problems and defects.

The basic architecture for a world-class quality system is a defect map, which is a flowchart of each product or product family showing each step in the process and identifying each defect opportunity—that is, each place where a defect can be created that would harm end quality. For all practical purposes, that means each critical specification. At each defect opportunity, the quality control in place is identified that assures the defect is caught before it passes on to the next step in the process.

The best quality controls are the simple ones that prevent the defect in the first place. That is where the Cpk studies and ensuring robust processes come in. Quality is almost guaranteeed if a very wide range of errors will not result in an unusable product. Better yet is the Japanese concept of *poka-yoke*, or mistake proofing. It is the idea of making it impossible to do anything other than the correct way. Engineering can make sure the wrong screw is never used by designing the product such that only one screw size is used anywhere. Everyone uses the right screw all the time because the right screw is the only one anywhere in the factory—or at least in an area of the factory. IT can eliminate data entry errors by building in edits making certain that only data in the right format can be entered in any particular field. You can't put the customer name in the address field because the address requires a combination of letters and numbers and the field will not accept anything with letters only, like a name.

The bottom line is that quality is the cornerstone of providing value to customers, and it is essential for cost-effective manufacturing. Excellent quality is a function of well-designed processes and practical quality controls. It requires a concerted effort on the part of everyone to drive cycle times down and to focus on process performance rather than inspectors wasting money checking other people's work. There is a great deal of substance underlying Six Sigma and ISO, but these concepts are also misused and can create the illusion of quality and consume a lot of money unless managers keep their eyes on those core principles. No points are awarded for passing ISO audits or for having more black belts on the payroll than the competition. The points are awarded only for excellent quality.

16

Sales, Operations, and Financial Planning

Of all of the tried and true management practices of the past, annual budgeting might be the most useless and perhaps the most destructive. In many companies the budgeting process is so complicated and fraught with manipulation that a sales forecast from as early as October for the following year is needed to have time for all of the number crunching back and forth required to lock in on a plan. Quite often the revenue plan is outdated before the budget year even begins. Undeterred, management marches through the year comparing everything that occurs to the fairy tale year the outdated budget describes. Accounting compensates with a blizzard of variance analyses—calculating everything from volume and mix variances to spending variances—all gauging the difference between reality and what has evolved into an utterly meaningless set of numbers.

Money may or may not be spent in September because it is or is not in a budget that departed from reality long before. People are promoted and given raises as a result of their ability to control spending against a pointless set of numbers rather than making decisions about how dollars are best spent to optimize the current realities of the business. Yet, in spite of the patent silliness of it all, every fourth quarter accounting moves to the front and center, and managers everywhere are consumed with budgeting work.

Of course, even in the unlikely event that management was able to accurately predict the market for an entire year, that knowledge was rarely built into the budget. In most companies budgeting is more a grand exercise in gaming the performance metrics than anything else. Sales understates expected demand to make it easier to hit the targets, while operations overstates needed spending to lighten the load for the coming year. Senior management usually knows that none of the budgeting input is credible and arbitrarily adjusts all of the numbers to compensate for the game playing as well as to create a fairy tale of its own—a plan for a successful year whether it is practical.

A static plan for the year is an exercise in futility mostly because there is nothing static about manufacturing and not much that naturally occurs in annual buckets. The only convenience is for accounting, which has to file tax returns and other financial and legal documents annually. As far as the business is concerned, there is no practical reason to expect January 1 to be anything other than the tick on the second hand following 11:59:59 on December 31. Manufacturing is a continuum, defined only by product life cycles that are not slaves to a calendar unless management forces them to be.

The alternative, and the much more powerful process for planning and executing, is a shorter-term approach called sales, operations, and financial planning (SOFP). Usually a rolling three-month outlook, SOFP is a process for continually ensuring that sales and operations are linked together in a cohesive manner to deliver on time in the most profitable manner possible. That three-month horizon may be longer in companies that manufacture longer lead-time products (e.g., capital equipment) but is rarely shorter.

Because the planning horizon is much shorter than an annual plan and because SOFP is a rolling, monthly process, the sales projections are more a plan than a forecast. Typically the sales leadership in the value stream brings a customer-by-customer, product-by-product plan for shipments over the upcoming 90 days based on the current realities with each customer. The sensitivities of the sales plan are introduced to the entire value-stream team and are discussed: what customers might order more or less of and why. The particulars in each customer's business are discussed, including what is happening that could cause unusual changes in the normal pattern of demand.

On the operations side of the discussion, because all of the expenses are assumed to be fixed, the starting assumption is that the actual levels of spending over the last few months will continue or be reduced over the upcoming few months. Reasons the fixed-spending plan was violated in the last month are discussed, and plans for cost reduction in the upcoming months are cross-functionally covered. Constraints—both real and potential—are identified and plans are set to manage them.

By linking production and sales together, taking into account the potential variations in customer orders, an inventory plan comes out of the SOFP meeting that assures customer deliveries will be made and that value-stream inventory objectives will be met.

In a broad sense, SOFP is a method of assuring that all participants in the value stream are linked together in a short-term plan that continually improves execution to their key performance metrics. The meeting ends with a consensus across every function within the value stream concerning the actions everyone will take in the coming few weeks to keep on track. The firefighting that occurs in most organizations as the functional departments are continually surprised by the activities in other functions—operations being hit by unexpected variations in demand, sales being surprised by production capacity shortages, months ending with wide variations in ending inventories from the beginning of the month—is largely eliminated. Who has to spend an extra dollar to save two elsewhere is well known and planned for. The drivers of variances are commonly understood, and the potential problems are known by everyone before they happen so that contingencies can be planned.

The SOFP process is also senior management's opportunity to monitor value-stream execution and performance. The senior folks typically sit in on the SOFP meetings and have a chance to see how things are going, such as how well teams are meeting their key performance indicators (KPIs), and to gauge whether things are under control. Because the role of the value stream is primarily to focus on continuous short-term planning and execution while senior management focuses on the longer-term, SOFP is senior management's chance to communicate longer-range issues and to ensure that the value stream is constantly positioning itself to be prepared for what it will encounter down the road.

A high-priority function of the SOFP process is capacity planning. Whether the horizon is 90 days or longer, the value-stream team is constantly looking at the workload on constraint resources—which bottleneck machines or people are beginning to reach levels of maximum use. The value stream gains a clear understanding of when and where pinch points will occur that could impact flow and customer service in ample time to head them off at the pass.

The importance of this ongoing planning, execution, and replanning process cannot be overstated. Although SOFP is structured around a financial review of sales and operations, it is much more than a traditional budget review session. For one thing, the nonfinancial KPIs are tracked and discussed. In addition, the process serves as primarily a communications vehicle that assures no one is retreating into silos within the value streams. It gives structure to make certain salespeople are closely attuned to operational issues, and that operations people know very well the challenges and opportunities the ongoing chatter between salespeople and customers has identified. An effective value-stream manager uses SOFP as the forum for making sure the entire value-stream team is focused on the fundamental mission of providing superior value for customers and that employees are not straying off on a path of their own. SOFP rallies the troops and reenforces the sharp focus on customers, processes, and profits.

Figure 16.1 shows a sample SOFP report format, indicating the last three months' actual results and the upcoming three months' plan. Previously we discussed the implications of calling just about all costs fixed, and it is here that the benefits of that approach come to the forefront. The objective is to assure that all indirect costs come in at the same level or below last month's levels. To the extent that there may be a variable element of any given cost, this will be difficult to accomplish as volumes increase. However, the rising costs are quite visible in the SOFP format and are the focus of discussion and attention. The value-stream team can focus on why indirect, usually non-value-adding costs have increased and can decide on a course of action to stop it. In traditional budgeting and cost reviews, those costs would be buried with allowable variable costs and would be far less visible to management.

By taking overhead costs out of the product costs and managing them as they really are—precise dollar amounts spent on specific activities—and by seeing them in the proper period, they can be managed and reduced. In the previous example, direct labor and materials comprise over 60% of the product, whereas costs such as depreciation and building maintenance each make up less than 2%. Looking at product costs in a typical standard cost format, they are trivial. Even if they were to be eliminated entirely they would not appear to have more than a negligible impact on the product, so no one focuses on them. In the SOFP format, they are seen for what they are—non-value-adding expenses that detract from profits—and any reduction that can be made in them results directly in an improvement in profits.

Kaizen events and Six Sigma projects are targeted efforts to direct resources at making rapid, cross-function improvements. An outcome of the SOFP process is the priority list for such efforts. Non-value-adding

VS 3 SOFP	LAST 3 MONTHS ACTUAL			NEXT 3 MONTHS PROJECTED		
	AUGUST	SEPTEMBER	OCTOBER	NOVEMBER	DECEMBER	JANUARY
SALES	$ 55,426	$ 61,307	$ 59,000	$ 60,000	$ 62,000	$ 56,000
DIRECT MATERIAL	$ 22,170	$ 24,523	$ 23,600	$ 24,000	$ 24,800	$ 22,400
Gross Contribution	$ 35,000	$ 35,000	$ 35,000	$ 35,000	$ 35,000	$ 35,000
DIRECT LABOR	$ 3,720	$ 3,950	$ 3,875	$ 3,800	$ 3,800	$ 3,800
Labor Associated Costs	$ 1,880	$ 2,010	$ 1,938	$ 1,900	$ 1,900	$ 1,900
Shop Supplies	$ 1,275	$ 1,300	$ 1,250	$ 1,200	$ 1,150	$ 1,100
Machine Breakdown Maint	$ 1,345	$ 995	$ 1,250	$ 1,250	$ 1,200	$ 1,200
Utilities	$ 2,222	$ 2,333	$ 2,444	$ 2,200	$ 2,300	$ 2,400
Miscellaneous	$ 2,345	$ 2,610	$ 2,400	$ 1,650	$ 1,200	$ 1,200
Management	$ 4,167	$ 4,167	$ 4,167	$ 4,167	$ 4,167	$ 4,167
Supervision	$ 3,333	$ 3,333	$ 3,333	$ 3,333	$ 3,333	$ 3,333
Depreciation	$ 833	$ 833	$ 833	$ 833	$ 833	$ 833
Building Maintenance	$ 1,100	$ 800	$ 850	$ 800	$ 1,300	$ 800
Miscellaneous	$ 150	$ 500	$ 300	$ 250	$ 250	$ 250
TOTAL OPERATING EXPENSES	$ 22,370	$ 22,831	$ 22,640	$ 21,383	$ 21,433	$ 20,983
GROSS MARGIN	$ 12,630	$ 12,169	$ 12,360	$ 13,617	$ 13,567	$ 14,017
	23%	20%	21%	23%	22%	25%

FIGURE 16.1
SOFP format.

costs that are not on a steady reduction track are high-priority targets for such improvements. SOFP is the ongoing tool to identify cost-improvement opportunities and to gain consensus for bringing all of the right people together to turn those opportunities into reality.

While the need for constant communication is always there and the work of managing the value streams is continuous, SOFP is the vital core of excellent management: bringing senior management and value-stream management together on a routine basis to continually focus on planning, execution, and improvement and to assure that everyone stays focused on the goals and values of the business.

17

Pricing to Win

The typical approach to sales and marketing has been to sell as much as possible of anything the company can make to whoever will buy them. More is better, so it is believed, and any increase in the top line is perceived to be good. Growth in sales year after year and quarter after quarter has become ingrained in business thinking as a critical measure of business performance, and the company with the greatest increase is the best.

In support of this, manufacturing and the often long tails of the supply chain are expected to follow sales up the peaks and down into the valleys as best as they possibly can. Companies are hiring as many folks as they can find and working overtime one quarter and are laying off and idling machines and entire shifts in another quarter. The result is a cost structure that is never optimized.

If we consider the companies rapidly emerging as best able to succeed in these dynamic times and attempt to boil down their approach to management into one overarching principle—and the one unifying principle behind every chapter in this book—it is all about managing the whole. Whether it is through functional organizations, standard costs, detailed performance metrics, or sophisticated enterprise resource planning (ERP) systems, the great failing of manufacturing management in the past has been to try to break the business down into small pieces and then to seek to optimize each piece separately. In doing so most companies have lost the ability to focus the entire business on the most important central objectives; they have lost the need for each person and each department to be mutually supportive in pursuit of those goals. Nowhere is this more destructive than in the gap between the sales and marketing efforts and the production efforts. It is also true that no greater improvement can be made to most companies than to bring these efforts together.

Figure 17.1 shows two potential sales curves over a 10-year period. While both companies enjoy the same overall sales revenue, the sales represented by the dotted line drive a constant state of capacity overuse, which drives overtime costs, delays in preventive maintenance, productivity lost due to hiring and training, expediting charges for parts and materials, then capacity underuse resulting in the costs of laying people off and leaving the benefit of their fixed-cost base on the table. What is happening is that the rate of flow through the factory is being continually changed, shoving more through it than it can handle and then starving it.

Conversely, the sales represented by the solid line with a fairly steady slope keep the factory at a constant state of capacity use, gradually but steadily increasing it as the plant breaks through constraints and finds productivity improvements. The cost structure of the factory—and the entire business—is always near its optimum state.

During Toyota's boom years, while most of the world focused on its manufacturing techniques—Lean *manufacturing*, the Toyota *Production System*—what flew under most radars was that its sales curve looked a lot like that steady solid line while its unprofitable American competitors had

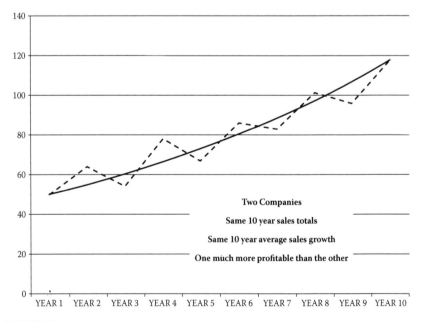

FIGURE 17.1
Sales curve comparisons.

sales patterns much like the wildly varying dotted line. Toyota's recent problems stem largely from its failure to maintain this strategy: pushing the slope of the sales curve up too hard and worrying too much about being the biggest instead of being the best. For 50 years they demonstrated that the tortoise's strategy was far superior to that of the hare in the race for long-term business success.

The big culprit in this, making it all possible and seemingly logical, is standard costing and the widespread approach to pricing that boils down to marking up the standard costs—looking to get a planned margin over standard costs on each sale. What makes this problematic is that standard costs include all of the overhead expenses, which as we have discussed are far more fixed than variable. The sales force sets prices based on a terribly flawed assumption that the fixed cost per widget can be determined and that the resulting cost holds true whether it is the first widget being sold or the millionth.

First, the fixed cost per widget can never be calculated simply because the fixed costs likely had nothing to do with any particular widget. Just as critical, the amount of fixed cost the selling price of the widget has to cover is completely dependent on whether it is the first widget or the millionth.

Pricing is not a simple mathematical exercise, easily delegated to someone with basic clerical skills or the product of math left to a pricing module in the computer. Pricing is a very strategic endeavor, involving the highest cross-functional leaders in the organization. Toyota achieved that steady slope by using pricing as the valve to be turned up or down as needed to assure a steady slope. Most of the excellent companies cited in this book do the same. They seek a planned volume level and view significant variation from that rate of sales—either up or down—as failure.

The gulf between selling and producing has to be crossed with a solid bridge. In fragmented organizations, decisions to outsource production— move it all to China—make sense directly as a result of the separation of the two functions. Since selling and making are organizationally and financially distinct, it doesn't seem to matter much where either one is done relative to the other. The gaping hole between sales and manufacturing in many companies is an opening companies such as Buck Knives and Wahl Clipper exploit when they charge in with a unified business and a strategy to provide superior value through their manufacturing capability, but it takes a tightly connected selling and manufacturing effort and a sound business strategy to make it happen.

Previously we discussed restructuring the business into value streams, each including all of the functional resources—including both sales and manufacturing—necessary to provide maximum value to customers. Each value stream is aimed at a distinct market, or channel within a market. In some cases a value stream can even exist to support one customer when that customer is big enough or important enough to warrant such singular focus. Each value stream, then, needs to be operating toward a specific strategy.

The critical inputs to forming the strategy include both market and capacity inputs. From the market side, you have to know the size of the market and your share of it, who the competition is, and how you stack up in the core value areas (i.e., quality, functionality, and reliability) as well as price. Most importantly, you must have a keen sense of how the customers define value, what their needs are, and the value stream's capacity.

The important discussion must be around the volume you need to attain to optimize the business. This comes from looking at the volume needed to prudently optimize the cost structure—typically at about 85–90% of capacity (not 100% because even Toyota needed to have some "wiggle room" to meet unexpected spikes in demand or to be able to recover from a disruption in production or the supply chain). It also requires knowing how much volume growth is available in the channel or market: do you currently enjoy a 90% market share or a 10% share?

In the event the company is near its maximum capacity but has a fairly small market share, how to install additional capacity is the obvious first step, and then the strategy has to revolve around determining the volume level to optimize the new capacity levels.

This exercise must take place uniquely within each value stream and for each market or channel the company serves. It is typical within the excellent companies for one value stream to seek significant volume growth and going to market with very aggressive pricing to do so, and another value stream to take a more conservative pricing and growth approach. This merely reflects the fact that one value stream has quite a bit of underused capacity, whereas the other one doesn't. For most companies, however, the objective is to grow sales and take advantage of idle capacity—especially during the current economic downturn. The task, then, is to learn how to use strategic pricing, rather than standard cost plus a markup, as the control valve to drive flow through the factory at the level everyone agrees will optimize capacity use (and therefore best leverage fixed costs) and will result in the market position desired.

The widely used approach to pricing begins with unit costing. While companies may use different allocation logic (e.g., assigning overheads on the basis of direct labor dollars or hours, machine hours, or multiple drivers as suggested by the activity-based costing school of thought), one way or another they all arrive at the same place; overhead costs are assigned to the various products. In Figure 17.2, a planned volume of sales is multiplied by the standard direct labor cost per product to arrive at a total planned labor expense. This total is then compared with the total planned overhead expenses, including both fixed and somewhat variable expenses to arrive at an overhead rate—400% in this simple example.

The total standard cost, then, is the sum of the direct material cost for each product, the standard direct labor, and overhead at 400% of the labor cost. While most companies take a more analytical approach, the result is the same: overhead costs are built into a "total cost" for each product. The pricing effort is then supposed to assure that each item is sold at a price significantly higher than the total cost to cover the sales, general, and administrative (SG&A) costs and leaves an acceptable profit.

If you follow the logic, you can see in Figure 17.3 that the "Total Standard Cost" is built on an assumption of volumes. Since it includes costs that are clearly fixed (e.g., machine depreciation and management salaries), using different volume levels would have yielded different total costs. The total costs are also blind to capacity use. The less capacity is employed, the lower the volumes used to calculate the cost will result in higher unit

	Unit Labor	Forecast Volume		Total Labor
Product A	$ 3.00	2,000	$	6,000
Product B	$ 3.50	1,800	$	6,300
Product C	$ 4.25	1,600	$	6,800
Product D	$ 5.25	1,400	$	7,350
Product E	$ 6.30	1,000	$	6,300
Product F	$ 7.25	1,000	$	7,250
		TOTAL	$	40,000
		Total Overhead	$	160,000
		Overhead Rate		400%

FIGURE 17.2
Sample overhead allocation method.

			Total Standard
Material	Labor	Overhead	Cost
$ 10.00	$ 3.00	$ 12.00	$ 25.00
$ 15.00	$ 3.50	$ 14.00	$ 32.50
$ 20.00	$ 4.25	$ 17.00	$ 41.25
$ 25.00	$ 5.25	$ 21.00	$ 51.25
$ 30.00	$ 6.30	$ 25.20	$ 61.50
$ 35.00	$ 7.25	$ 29.00	$ 71.25

FIGURE 17.3
Sample standard costs.

costs. If a machine has a $1,000 depreciation expense, for instance, and is capable of producing 10,000 units per year, the depreciation rate would ideally be 10¢ per unit produced. However, if the factory is only utilized at a 60% rate—6,000 units to be made by the machine; then a depreciation cost of 17¢ will be built into the standard cost. The point is that the lower the volumes, the higher the costs. This is just a matter of common sense, some might say, but many companies go into a "death spiral" as a result of this phenomenon. Volumes are down for whatever reason, causing costs to increase, leading to higher prices and then fewer sales and then further cost increase and so on, until outsourcing production to a country with low labor costs seems to be the only solution to their high costs.

At the heart of the problem is that fixed overheads are not the cost to produce any particular unit or even any particular volume. They are the costs of capacity. It is also true that most companies have overhead rates far higher than the 400% in our simple example. Due to automation, mechanization, and a long focus on basic productivity improvement, most companies have very low direct labor costs (often 5–6% of sales) but increasing overhead costs. The more costs shift to capacity costs and away from true unit costs, the more destructive standard costing and the pricing decisions that flow from it become.

Using the standard costs from our previous example, the typical analysis yields something like the summary in Figure 17.4. Without really understanding the implications of the numbers, one would look at Products E and F and come to the conclusion that a serious problem exists with the cost–price relationship. E contributes only 2% to SG&A and profits, whereas F is "losing money," or is underwater, with each sale.

	Price	Standard Cost	Margin		Planned Volume	Gross Margin
Product A	$ 30.00	$ 25.00	$	5.00 17%	2,000	$ 10,000
Product B	$ 38.00	$ 32.50	$	5.50 14%	1,800	$ 9,900
Product C	$ 48.00	$ 41.25	$	6.75 14%	1,600	$ 10,800
Product D	$ 62.50	$ 51.25	$	11.25 18%	1,400	$ 15,750
Product E	$ 63.00	$ 61.50	$	1.50 2%	1,000	$ 1,500
Product F	$ 70.00	$ 71.25	$	(1.25) −2%	1,000	$ (1,250)
					Total	$ 46,700
					SG&A	$ (40,000)
					Profit	$ 6,700

FIGURE 17.4
Typical unit "profitability" calculation.

The logical next steps would be to try to raise the price, to outsource production to some lower cost source, or to discontinue selling the item all together. None of these solutions are likely to be very effective, and since they are based on flawed logic they are likely to be wholly unnecessary to begin with.

If the price could be raised without losing sales, in all likelihood, the sales and marketing folks would have done it long ago. More often, the product is repackaged or given a superficial facelift, or advertising money is thrown at it in an attempt to drive a price that is truthfully out of line with the basic value of the product. Although sometimes it can make sense, raising prices is rarely a viable option. In fact, more often the company is facing pressure from customers and competitors to reduce prices. Any marketing or brand management solutions aimed at increasing prices relative to the real value of the product are bound to fail eventually, if not right from the start.

Outsourcing production and discontinuing production are even more destructive. Products E and F have a combined $54,200 of overhead assigned to them, most of it fixed. When you farm them out or simply walk away from their volume, most of that $54,200 stays with you. It will end up being allocated to the remaining products, driving their "total cost" up and detracting from your profits and inevitably causing you to think you have to get higher prices or come up with other solutions for the rest of the products. This is another path to the "death spiral" that only can end up in disaster.

The much more effective way—strategic pricing and strategic cost management—begins with rolling the numbers up a bit differently. As the following example indicates, the numbers are the same as the previous charts with one major exception. We have pulled the allocated overheads out of the unit costs. Note that the companies committed to the principles of lifetime employment—those that have truly built the cultural ideals of the first section of this book into their business model—also exclude labor from the product cost math. By committing to employees, they become a fixed cost too. They are another capacity cost. It is precisely through this logic and thought process that Toyota has avoided layoffs for more than 50 years and that Wahl Clipper has gone 35 years without a layoff. While most companies, as much as they would like to commit to their employees, struggle with the idea because it just doesn't make sense to them financially. The chart in Figure 17.5 reveals the beginning of the different set of numbers the successful companies use. Looking at things this way, committing to employees makes perfect financial sense. This view of the numbers demonstrates that every product makes a contribution to the combined need to cover SG&A, profits, and the overheads truly related to capacity, rather than varying volumes and mixes of production.

The next step is really quite simple: Play "what if" games with pricing and volumes. This entails applying the volumes targeted in the strategic session and the prices necessary to achieve those volumes. In Figure 17.6, some

	Price	Direct Material & Labor	Planned Volume	Gross Margin	
Product A	$ 30.00	$ 13.00	2,000	$	34,000
Product B	$ 38.00	$ 18.50	1,800	$	35,100
Product C	$ 48.00	$ 24.25	1,600	$	38,000
Product D	$ 62.50	$ 30.25	1,400	$	45,150
Product E	$ 63.00	$ 36.30	1,000	$	26,700
Product F	$ 70.00	$ 42.25	1,000	$	27,750
			Total	$	206,700
			SG&A	$	(40,000)
			Overhead	$	(160,000)
			Profit	$	6,700

FIGURE 17.5
Margin analysis using direct costs only.

prices actually went up, with a corresponding decrease in volume, whereas most went down but generated a disproportionate increase in volume.

The important ingredient is the knowledge of where the value stream stands from a capacity use standpoint: how much additional volume can the value stream produce without a significant increase in the fixed cost base needed to support the capacity? The goal is to find the point that drives the "right" volume through the value stream that hits the optimum total cost structure and fulfills strategic sales objectives.

Note in Figure 17.6 that the prices identified by most of the truly outstanding companies we have cited are the prices needed to undercut their standard cost-driven competitors. It is easy to see how a company deploying this approach to pricing will have an overwhelming advantage over another trying to recover allocated fixed costs on each unit with each sale.

Of course there is always the possibility that this approach to pricing will yield prices necessary to meet volume requirements but will not generate a contribution to profit for the value stream. It is important to note that this is the truth, which is usually hidden in standard costing's general allocations and broad-brush approach. The truth has to be confronted, and that is where the SOFP process has such potential.

The costs that have to be reduced are not necessarily related to any particular product. Rather, the overall cost of the value stream has to be minimized. The direct link between reducing the non-value-adding overhead

	Price	Direct Material & Labor	Planned Volume	Gross Margin	
Product A	$ 29.00	$ 13.00	2,300	$	36,800
Product B	$ 39.00	$ 18.50	1,650	$	33,825
Product C	$ 48.00	$ 24.25	1,600	$	38,000
Product D	$ 64.00	$ 30.25	1,300	$	43,875
Product E	$ 62.00	$ 36.30	1,100	$	28,270
Product F	$ 68.00	$ 42.25	1,200	$	30,900
			Total	$	211,670
			SG&A	$	(40,000)
			Overhead	$	(160,000)
			Profit	$	11,670

FIGURE 17.6
Margin analysis using different pricing scenario.

costs (e.g., salaries and factory support labor) and profit objectives can be clearly identified, and cost-reduction efforts can be launched to meet the strategic objectives.

The point to take away from this chapter is that pricing should be an ongoing effort, continually tying total operating costs with sales realities and customer requirements to ensure ongoing bottom-line success. This way of thinking is a radical departure from unit costing, unit pricing, and unit margins and dispels the notion that some products are profitable whereas others are not. In fact, only businesses can be profitable, and just about every product can make a contribution to profits in the right volumes.

This strategic approach to pricing and cost management is the financial engine that turns the focus on flow and capacity into extraordinary success. The keys to success are a unified, broad approach to the business by all of the senior management team, especially sales and marketing, manufacturing and supply chain, and the financial staff. All senior management and value-stream team members have to be continually engaged in broad planning and decision making and must look at the business as a whole.

18

Capital Investment

Everyone knows that you cannot buy happiness, but less broadly understood is that you cannot buy profitability either. By some estimates, General Motors under Roger Smith spent close to $30 billion on robots and factory automation as its competitive response to Toyota. It did not succeed, as could have been easily predicted, and served only to hasten General Motors' demise. Nothing for sale can give you a leg up on the other guy. If robots were the key to success, Toyota could have bought just as many of the same ones. Any machine you can buy, your competitors can buy too. Any factory you can build, so can they. In the end, success or failure is a function of management. The winner is always the company that manages best, not the company that spent the most. The downward spiral of American manufacturing is littered with poorly managed companies that thought they could solve their problems through the acquisition of successful small companies; this only destroyed them and the companies they bought.

Capital spending should be very judicious and operate as an extension of good management, not a replacement for it. One of the authors was profoundly struck several years ago by the sight of a state-of-the-art electrical discharge machine in a Chinese factory nearly adjacent to a 1950s vintage stamping press. It was a vivid illustration of using the most appropriate technology for the job, not the highest level of technology but not the cheapest either. That principle of investing in the most appropriate technology—neither the latest or the cheapest—for the work you are trying to do should be the guiding principle behind capital spending.

The major failing in most capital spending decisions is the use of some variation on ROI logic as the basis for the decision. Such analyses are based on largely unknown and unknowable financial considerations and are inevitably tilted toward labor cost reductions simply because labor cost is the only input most companies find easy to quantify. Capital spending often becomes little more than a hunt to find machines to replace people. One of the more obvious insights from Japanese companies is that the labor savings from investments in machines rarely happens and that most often you end up replacing lower-paid production people with more highly paid specialists to take care of and set up the expensive machines and to deal with the quality problems machines often compound. As the Motorola originators of Six Sigma were fond of saying, with automation you can create defects at a rate you never before thought possible.

The most profound application of the head, heart, and gut approach to decisions is in the area of capital spending. Of course the cost of the machines should be considered, and whether it is going to reduce or increase costs should be determined as best it can. But investments primarily aimed at eliminating people fail the heart test, and many investments utterly fail the commonsense gut test.

In considering the numbers, as with metrics, the numbers that matter are the bottom-line outputs. As best you can, a current-state and future-state statement of value-stream contribution to cash flow, inventories, and profits should drive the numerical analysis. This requires a keen understanding of the constraints in the overall process and an appreciation for whether the investment is an improvement at the bottleneck or one of Eli Goldratt's "mirages" of improvement somewhere else.[26] The primary impact capital investment has on the economics of a value stream is the impact on flow. If the new equipment will facilitate faster end-to-end flow it is likely to be a very good idea; if not, an eyebrow should be raised at just why the equipment is being purchased. Eliminating labor cost through the use of a machine that will crank out parts faster than the upstream or downstream operations can process them or faster than customers want them is almost always a false economy, no matter what the ROI numbers say. The two most powerful justifications for investing in new machines are (1) to break through a constraint in factory flow and thereby leverage the fixed costs of the factory, and (2) to increase the capability of a process, or to widen the narrowest garage door, so to speak. The next best basis for

---

[26] Goldratt, E.M.; *The Goal: A Process of Ongoing Improvement*; Gower; Surrey, UK, 1993; for additional information on the Theory of Constraints and Throughput Accounting see the Goldratt Institute Web site http://www.goldratt.com/

investment tends to be to improve changeovers by reducing machine setup times. Way down at the bottom of the list of investments that truly pay off is one that reduces labor cost.

The best investments almost always improve quality and flexibility. They can turn on a dime and start to produce a different part in a very reliable manner. These things are tough to quantify in a traditional accounting system, however, and cause too many companies managed by the numbers to miss the boat of opportunity. It is important to point out that, while success cannot be bought and management is the driver of success, management is typically manifested in its decisions concerning what and what not to buy. Poorly managed companies spend a lot of money on machines to cut labor costs. Well-managed companies spend their money on machines to blow production from one end of the factory to the other faster and with higher precision. It is the difference between night and day, but because traditional accounting systems support the former approach, the companies that hold onto old financial ideas suffer.

Common sense and a keen understanding of the value proposition the company provides must be big drivers of capital spending. In the excellent companies there is a very low threshold for investing in technology that improves the core of the product; for example, machines that will result in a better clipper blade at Wahl or a better-sealed window at Andersen find a pretty easy path through the approval process. Machines that improve peripheral products or features find it tougher going. Investments in non-value-adding tasks, such as computer software for management, often find approval virtually impossible regardless of the ROI. Investments that expand the capacity to produce core products similarly get an almost free pass.

19

Performance Metrics

In the wildly successful movie *Rain Man,* Dustin Hoffman played a character who was an autistic savant. He had an amazing ability to calculate, memorize, and spit out detailed information but absolutely had no idea how to put it into perspective. Living in the information age for the last 30 years or so has made managerial savants of many when it comes to metrics. We have access to prodigious amounts of data, and the fact that like the "Rain Man," we often have no concept of perspective does not stop us from trying to act on it.

Rain Man committed the phone book to memory but had no idea how to make a phone call. He memorized every statistic on the backs of thousands of baseball cards but had never been to a baseball game. In like fashion, managers track and recite statistics on headcount, labor efficiency, and points of product sales margins and can often provide financial ratios to two decimal points; all the while, the company is losing money.

The first half of Albert Einstein's quote, "Not everything that can be counted counts," applies directly to this obsession with metrics. Just because someone can gather the data and put it on a chart or can compare it with some other number and calculate a ratio does not mean it is important or that it has any meaning at all. It certainly does not mean that management team members should form any conclusions or take action simply because they have data.

Perhaps the worst abuse of information simply because we can get it is to attempt to boil our assessment of people's performance down to some number we have rather arbitrarily chosen to call a metric. This applies to the second half of Einstein's quote: "Not everything that counts can be counted." No one's contribution—or lack thereof—to the company can be summed up mathematically, no matter how much we would like it to be so or no matter how much easier that would make the job of managing people.

The problem is that we confuse measures of performance (i.e., bottom-line outputs and results) with subordinate bits of data that may or may not have an impact on those results. We also suffer from a limited view and a limited understanding of all of the different variables that interact along the way to achieving the results. Labor efficiency, for instance, is important only to the extent that it provides one small bit of information concerning total costs. By itself, it is neither good or bad; it is meaningless. The stockholders could care less about labor efficiency, although they hope members of management are on top of it and are using it as one input of many into the process of achieving overall profits.

Manufacturing is a lot like baseball when it comes to metrics versus data. Baseball fans can become obsessed with statistics, but all that really matters is winning games. Earned run averages and on-base percentages mean nothing if the team is in last place. They also mean nothing when the team is in first place. They are simply data points a good coach uses— along with a lot of other data points, judgment, and experience—when making the decisions.

Many of the subordinate measures of how processes are performing— and that is what they are—can be traded against each other, which is why management must be very careful with them. High labor efficiency is a good thing to have, but not if it comes at the cost of higher inventories or poorer quality. Inventory reduction is generally good too, but not if customer delivery suffers as a result. Therein lies the problem with using these data points as the basis for making decisions or thinking they are valid measures of performance. Unless management team members know all of the possible side effects of emphasizing a particular aspect of the performance of a process, they are opening themselves up to disastrous unintended consequences when they push one metric too hard to save a dollar, and end up spending two dollars in some other area.

The only real measures of performance of a process is the absolute output. View the factory—or at least a value stream within the factory—as a "black box." All you know for sure is what went in and what comes out. Those absolute inputs and outputs are the only basis for valid performance metrics. While each company has to decide for itself what to measure, the inputs are basically the money that went in, and the outputs are profits, cash flow, customer delivery, and quality. These bottom-line results compared with resources that went into the black box are what really matter. Everything inside the box is simply data that enable management to better understand and affect the real outputs.

Headcount, labor efficiency and productivity, overall equipment effectiveness, quality inspections at some interim step in production, floor space used, and machine use are all bits of data helpful only if they provide insight into improving profit and cash flow.

When the company was compartmentalized by function, using these internal bits of data as substitutes for performance metrics may have been a necessary evil. However, when the company is aligned into value streams, everyone should have a clear line of sight to the bottom line. The measurement of a value stream and the metrics for the performance of the management teams are the bottom-line results of the value stream: its contribution to profit, its inventory levels, cash flow from the value stream, and its customer outputs, such as quality and delivery. In devising schemes to pay bonuses or create an incentive for the value-stream management, these bottom-line output metrics should be the only ones used. Who cares what overall equipment effectiveness (OEE) or the direct to indirect labor ratio was within the value stream if it is increasing profit contribution and cash flow, customer deliveries, and quality? Conversely, why would we want to pay someone to improve those internal metrics if the value stream is not performing well in its outputs?

It is very important to segregate the few things that really matter and measure them diligently. It is also important to measure them as often as practical and measure graphically. Fixed-point measurements are dangerous, such as quarterly profit contribution compared with last quarter or to the same quarter last year. Such measures create a powerful incentive to take harmful short-term actions, as everyone who has worked in a factory belonging to a publicly traded company knows. The activities in such plants to make this quarter's numbers look good, regardless of the impact on next quarter, are often ridiculous. Plants irritate customers by shipping early and suppliers by refusing to take delivery of incoming materials just to hit a quarterly inventory target. Vital preventive maintenance and employee training are deferred for a few weeks simply to beat this quarter's cost targets.

If cash flow, for instance, can be measured weekly, all the better; if not, then monthly. The point is to capture the data in the most frequent intervals practical and measure them on a graph. Trends are much more informative than widely spaced data points.

It is important to keep in mind that, once data are used for measurement and once people are impacted by the results of those measurements, the data will cease to be objective. They will be influenced as much as possible by the people whose lives are affected by the results of the measurements. That is not necessarily a bad thing, but managers have to be very careful what they ask for, lest they get it. It is far better to ask for steadily increasing cash flow and profits over the long-term than to ask for a very specific "stretch goal" by a set point.

None of this means those subordinate bits of data are unimportant or should be ignored. By all means, senior management members should insist that value-stream management identify, track, and properly use all of the relevant indicators of how well their processes are performing. The point is that these data should not drive decision making unless they can be shown through a clear and direct line that the decisions will have a positive impact on those bottom-line, output metrics without trading one for another.

There is only one exception to this arm's length approach to subordinate data within the value-stream "black box": the use of OEE at the bottleneck operations. This should not be optional. We discussed at great length the importance of flow and the huge impact the performance the bottleneck operation has on overall flow: an hour saved at the bottleneck is an hour saved for the entire process and so forth. We also pointed out that there is always a constraint, though it may change. That means one operation is always the primary driver of overall flow through the processes and the controller of the overall costs. Applying OEE at the constraint should not be optional. This is no casual, technical, detailed shop-floor metric. It is right at the heart of manufacturing management because it embodies everything important: managing the constraint, optimizing flow, understanding that direct labor and machine use all by itself is destructive. When people ask why senior management should be involved in the details of a metric like this, the answer is because it is the acid test of whether management really understands manufacturing excellence. If OEE is perceived to be just another shop-floor detail best left to people a few rungs down the ladder while the senior folks focus on the important, heady issues of management, it is a clear sign that the senior folks have missed the boat.

OEE is simply a combination of machine availability, its use, and its yield. It is calculated as

$$A \times U \times Y = OEE$$

where

A = the amount of time a machine is actually available as a percentage of time it was planned to be available. In other words, if you plan to work two shifts and have the machine available for 80 hours minus 2 hours per week for planned maintenance, availability (A) would be the number of hours the machine was actually in working order and available to be used divided by 78. So if the machine is actually available to run only for 70 hours, then $A = 70 \div 78 = 90\%$.

U = the production of the machine compared with the actual requirements on the machine. Note that this is not necessarily the same thing as the maximum the machine can produce. Rather, if the true customer demand from the machine is 100 pieces per hour during the 78 hours the machine is expected to be available and the actual output is only 95 pieces per hour, then $U = 95 \div 100 = 95\%$. (The maximum value allowed is 100%; no credit is given for overproducing to the schedule.)

Y = the yield from the machine in terms of good pieces. If the machine produced 6,650 pieces and 133 of them were bad, then $Y = 6,517 \div 6650 = 98\%$.

So if $OEE = A \times U \times Y$, then $OEE = 90\% \times 95\% \times 98\%$, or 84%.

This is a very important composite measure of the effective management of a constraint resource and should always be employed rather than simple machine utilization. Where simple machine use measurements can encourage overproduction and poor quality, OEE is a balanced measure of all of the essential elements of bottleneck optimization. As good as it is for measuring the use of a critical resource, it is not a particularly helpful measure in nonconstraint resources. As with all of the subordinate data within the "black box," it is important to know which of the measurement tools to use when and how to use it.

Earlier we mentioned two very fundamental metrics of the value-driven, excellent companies: sales to value-adding costs (SV) and value adding ratio (VAR). These help boil down Henry Ford's adage that "profit is the inevitable result of work well done" to its core.

Taken in reverse order, VAR is a measure of how much money is being "wasted" on things that do not increase sales value. The objective of the company is not to reduce costs, contrary to conventional but uninformed wisdom. It is to spend as much of the money going out the door on things that will result in revenues. This pure entrepreneurial approach is the hallmark of an excellent company. The startup business wastes as little as possible on overheads, churning everything into products that can be sold. The mature company loses this sense of urgency and priority, becoming laden with administrative and bureaucratic expenses the entrepreneur would never dream of incurring.

How a company chooses to determine which expenses add value and which do not is purely a matter of understanding the products and the customers. Most direct materials are value adding, as is most direct labor. There are plenty of gray areas, however. Packaging is one. To the extent that the packaging protects and organizes the various elements of the product it is a value-adding cost. To the extent that it is advertising, it does not. Of course the packaging must clearly identify the product and assure the customer of exactly what is inside the package. For the most part it is safe to include all packaging costs in the value-adding cost calculation, but there are many cases in which marketing has gone over the top, spending as much or more on multicolored packaging to induce the customer to buy the product than was spent on the actual product. While this may add value to the salesperson's commission, it does not add value in the eyes of the customer.

Many machine-related costs are also in a gray area. Machine depreciation and preventive maintenance may well be value-adding costs, whereas breakdown expenses are not. Most of product engineering can be classified as adding value, except for the time product engineers spend correcting quality problems stemming from design flaws and their pure "blue sky" research and development (R&D) time.

Members of each management team must decide for themselves what adds value, but team members should spend considerable time exploring the decision. If nothing else, it makes for a very interesting and lively discussion as they form a consensus around what really matters. More importantly, the discussion will lead to a clarity of focus on what is important to the customers and, hence, to the company.

The other key measure, SV, keeps the first measure honest. It evaluatees the degree to which costs assumed to add value to the customer are being translated into sales dollars. While it can be seen as a measure of pricing effectiveness—is the sales and marketing effort commanding prices commensurate with the value being provided?—it is more often a measure of how well the company understands value-adding versus non-value-adding costs.

One company we know included all labor costs in the value-adding category because it did not want to offend any employees by implying that they were not valuable to the company. While its intentions are admirable, value adding is not a measure of people's worth. It is a measure of what customers are willing to pay for, and the company would have been better served correctly identifying its true value-adding work and finding another way to reassure its employees of their inherent worth to the company.

The value of SV is that, in the case of that misguided company just cited, adding more employees would have made its VAR look good, those mistakenly classified "value-adding costs" would not have translated into higher sales dollars, and the SV would have suffered.

In this manner the SV metric is a powerful tool for filtering cost increases in the value-adding area. If someone wants to do something in one of those gray areas (e.g., spend more on packaging), the criteria for the decision should largely revolve around whether that addition to packaging costs will directly translate into increased sales revenues.

Taken together, keeping a close eye on how the SV and VAR numbers improve is the best way to keep a value-driven company on track, profitable, and growing.

Concerning the SV and VAR metrics, most importantly these figures should be the subject of an ongoing discussion among all members of the senior management team. These numbers define the core of the business' value proposition and its basis for existence. A consensus understanding of exactly what creates value for customers—and what does not—is vital to ensuring that the entire organization is working in unison toward the same goal. There can be no major disagreements among the senior managers on this score.

20

Wrapping It Up

We have created a pretty lengthy to-do list for management—nothing short of a complete overhaul of how you think and how you run every aspect of the business. The good news is that it doesn't have to be all done at once. The transition from where you are to where you have to be to join the elite ranks of the companies defying conventional wisdom and succeed in countries where manufacturing cannot compete, in industries that can be served only by countries with low labor costs, is one that never ends. You simply have to change the way you think and then start to change the way you run the business.

The bad news is that this is not a cafeteria deal. You don't get to pick the elements of managing differently that are easiest to do and leave the rest alone. It is an all-or-none proposition. Either you believe that it is all about creating value for customers, or you don't, in which case you will continue to chase cost reduction for its own sake.

Either you believe that cash is real and is the only aspect of financial management that really matters in the end, or you don't, in which case you will continue chasing paper profits and optimizing financial ratios.

Either you are committed to a responsibility for all of the stakeholders and believe that a commitment to them will result in a commitment from them to you that will pay off in a big way, or you don't, in which case you see them all as "headcount" and "inputs" to the profit chase. You buy into that old Alfred Sloan theory that basically says the stakeholders buy their tickets and take their chances.

Either you have the courage to step up to the idea that the problems on the factory floor are a result of management and that management has to undergo radical change, or you don't, in which case you choose to believe in the infallibility of old management theories and opt to put all of the onus on manufacturing to perform better or die.

You really don't have much choice. The economic fiasco of the past few years has already created the watershed. The companies stuck in old management thinking are the ones in the most trouble and the ones at the head of the line at bankruptcy court, by and large. If you weren't already starting to question some of the management thinking—especially the thinking that said cash is just another commodity and inventory is just another liquid asset—then you are probably doing so now.

If nothing else we hope we demonstrated that simplicity and focus on commonsense fundamentals has power. Question everything, especially the complicated and expensive solutions. You have to get out of the spiral of pursuing complex solutions and then, when they don't work, compounding them with greater complexity. The enterprise resource planning (ERP) system isn't helping much, but you need to quit adding to it by implementing complicated forecasting systems when the root of it all is long lead times. Shorten the lead times, then get rid of the forecasting system, and then gut the ERP system down to its basic core: simple, uncomplicated.

Purge the accounting reports and financial analyses that nobody really understands, and just spend less cash. Start running the business more like you run your life, where your basic values, the wisdom from hard-earned experience, plays a bigger role than spreadsheets.

The best your company ever operated was in its infancy when there was no bureaucracy, few meetings, no wasted effort and money, when people did a lot of different jobs, and no one was a dispassionate professional thinking what you did could be boiled down to numbers and charts. People had energy and focused on customers and cash. Inventories were lean. The people who knew the most about the products were the people in charge. Implementing management processes was what drove it away from those high-energy, sharp-focus days. Simple excellence is simply a return to those days and that approach to managing.

Perhaps the best outcome from joining the companies achieving excellent results from simplicity is that the people working for those companies have more fun and a sense of fulfillment their bureaucratic competitors never know. People are more empowered: everyone knows what the core objectives are and believes he or she plays a part in meeting them. Value-stream structures put people from different backgrounds together for a common, important purpose, and a sense of teamwork emerges that people hunkered down in functional silos—all believing they are under-appreciated and misunderstood—never feel.

Knowing they are working for a company that shares their personal values gives people that sense of purpose and fulfillment Bob Chapman at Barry-Wehmiller sees as the most important aspect of leadership. It is simply an approach to business that puts the common sense and integrity of everyone, from the executive suite to the shop floor, at the forefront. Most importantly, it works and results in success many believe to be impossible.

Index